Rise Up
with Wings Like Eagles

Discover inner strength and wisdom to transform our relationship with the Earth

Rise Up
with Wings Like Eagles

Discover inner strength and wisdom to transform our relationship with the Earth

Chris Sunderland

EARTH BOOKS

Winchester, UK
Washington, USA

First published by Earth Books, 2016
Earth Books is an imprint of John Hunt Publishing Ltd., Laurel House, Station Approach, Alresford, Hants, SO24 9JH, UK
office1@jhpbooks.net
www.johnhuntpublishing.com
www.earth-books.net

For distributor details and how to order please visit the 'Ordering' section on our website.

ISBN: 978 1 78535 464 9
978 1 78535 465 6 (ebook)
Library of Congress Control Number: 2016936943

A CIP catalogue record for this book is available from the British Library.

Design: Stuart Davies

Printed and bound by CPI Group (UK) Ltd, Croydon, CR0 4YY, UK

CONTENTS

Introduction

Many people who have read early versions of this book have urged me to say more about myself, and to reveal more of what lies behind the text. In response I have laid aside the original 'very earnest' introduction to this book and written something more personal.

On the face of it, I have had a rather chequered career. I grew up in the very sheltered environment of a minor public school, attempting to survive emotionally and succeed in all the ways that success was described to me by the school. At the age of 14 we were asked to choose between science and arts. I chose science. After all, it seemed so much easier and more certain than studies like history. Looking back on it this decision was to shape my life.

I won a scholarship to Oxford, took a first in Biochemistry, then a doctorate, and became a research scientist, who was privileged to study in a world-class biochemistry lab in Oxford, where we were developing tools known as monoclonal antibodies that have since become commonplace in medical research. I moved on to apply the same techniques to the human placenta in an attempt to shed light on the mysteries of graft rejection as well as trying to develop reagents for detection and treatment of gynecological cancers.

I think others might judge that, at this stage, I was a fairly narrow-minded individual. Science came easily to me. The adventure of looking down a microscope each day and seeing things that no one had ever seen before was intoxicating. But it was such a small world. There were a few of us in my very specialized field and we would meet every six months or so at conferences, compare notes and go back to work. I was longing for something more.

There was also a tension within me that needed resolving. I

had become a Christian at university and was trying to work out the implications of that. The certainties of the scientific approach to truth that I had been brought up with were all that I really understood at this time and I initially tried to translate my scientific approach into a similar certainty in the faith realm. This made me prey to fundamentalists and extremists of various sorts and my wife and I joined, or nearly joined, some catastrophic communities! This experience prompted a deep questioning that has lived with me through the rest of my life and has matured to the sort of position you will see reflected in this book.

I am now committed both to the importance of the scientific method in researching truth relating to the material world, and also to the importance of other means of seeking truth in relation to human society and its working. Readers will notice this particularly in Chapter Three, where I set out a bold position about two types of truth, scientific truth and narrative-based truth. I have come to see that human beings use narrative continually in public conversations that seek truth in the social sphere and that narratives bundle up and express the enormously complex world of our deepest commitments, values and feelings so as to provide a clear basis for human action. I have held this view for more than ten years now, so it has stood the test of time, at least in my mind, and I am frustrated that other thinkers and the general public are so unclear are about these things. This is illustrated in Chapter Five by reference to public arguments about climate change.

My faith journey led me to offer for ordination in the Anglican Church and I was duly appointed vicar of a tower-block estate in inner-city Bristol in 1990. This was quite an experience for someone with such a sheltered upbringing. It was challenging socially, in terms of being with folk who had lived lives that were completely off my previous radar, in terms of the hardship and tragedy they had experienced and also the sheer strength of their community. It was also challenging in terms of faith. Rather than just being a Christian, responsible to myself for what I said about the faith, I was now part of the institution. It was hard enough trying to make sense of the faith itself to people but the Church, with all its tortuous, historical interpretations of how this or that ought to be done, provided another whole layer of challenge if one was to relate meaningfully to people.

So, there I was trying to draw alongside folk in one of the most challenging estates in the UK with a simple faith message, while working for this church which was entangled in a set of rules and procedures that were looking more out of touch by the day. I also lived in the one substantial house in the parish with my wife, growing family and aged stepmother, while many others lived in tower blocks. The contradictions in the role were evident from the outset and the lives of the people I met were a challenge to come to terms with, especially when set against my own privileged upbringing. I recall encountering a quality of goodness in people, which was often set in the context of an adversity and tragedy the like of which I had never previously encountered. The whole experience was life-changing for me. I tried to be of some help to those I met and some would say I was just that. Yet, overall my most profound impression was of a mismatch of expectations. Anglicanism simply was not working as it was designed to do. The culture had shifted so that the old establishment assumptions of faith could no longer be sustained and I was left with layer upon layer of justification that simply did not wash with the folk around me. There was also a considerable weight of grief bearing down on the community and it seemed necessary to shoulder this grief wherever I found it. It was probably this grief, above all else, that got to me in the end and I realized that I needed to leave for the sake of my own well-being. I resigned my post as vicar in 1998.

The one thing that had really burnt itself in on me in my life to that point, was that our society was losing the benefits of religion, but had nothing to replace it with. As a way into thinking about this, I wrote an MPhil on the subject of

'Individualism and the Breakdown of Community Life in Western Society' and began to run courses in pubs on themes like *Weapons of Mass Delusion* (referring to the Iraq War), *A Surveillance Society, Immigration – Threat or Opportunity?*. I was able to use narrative actively now, in the form of public discussions, to ask the primary questions of our society, as I saw them. Faith perspectives could take their part in these conversations, as a deep-level commitment and a way of seeing the world that could appeal as narrative without demanding allegiance. That was back in 2001 and I am shocked to see how relevant these questions remain today. It was all part of feeling my way towards a new manner of relating what I saw as the insights coming from my faith position to the post-Enlightenment society. In the process I founded a small charity known as Project Agora, named after the public space in the center of a Greek city, which was about active citizenship and provoking conversation on important issues. Agora, from the outset, was committed to nurturing social change without attracting attention to itself and it has quietly supported many of the initiatives I have been involved with in the following years.

It was around this time that I laid the foundations for thinking that are exemplified by the early chapters of this book, such as the inadequacy of simple rationalism and the need to incorporate emotions, deep-level commitments, drives and empathy into our public decision-making. I also began to see society as a trust and recognized the importance of trustworthiness, challenge and creativity. I have lived and worked with these great themes for many years now and am sure of their value. This book is a synthesis of that work and I hope that the earlier chapters will offer some philosophical and emotional clarity and confidence to many people who work for change today.

This whole process of change in my own perspectives seemed to pave the way for the shocking revelation about our relationship with the planet that hit me with full force around

2005. Suddenly I realized that our relationship with the Earth would be the big narrative of the twenty-first century and that solving this was possibly the greatest challenge that human beings had ever faced. Many others had seen this before, but when this truth finally dawned on me, it came as a sort of epiphany. I had often wondered what my life had achieved up to that point. In my worst moments I would fear that I had achieved little as I had hopped from one career to the next without ever really accomplishing anything obviously significant. Yet now I could think of nothing better to do with the rest of my life than dedicating myself to addressing the challenge of finding a way to live in harmony with the Earth.

I hoped, of course, that the Church would embrace this new challenge and formed an initiative known as EarthAbbey to encourage a process of living more in harmony with the Earth from a faith perspective, but, sadly, I have to report that the Church has not responded with anything like the interest or vigor that I had hoped. Chapter Eight presents my reflections on why this might be the case.

Yet that is not the end of the story. I happen to have lived, for the last thirty years, in an extraordinary city. Bristol, UK won the accolade of European Green Capital for 2015, not because it is already very green. It is not. It has a dreadful transport system, for a start, with shocking levels of air pollution. No, it won the title because it is a seed bed for creativity in green initiatives and has a community of hundreds of people looking for different ways of living, some of whom are living on very little money in order to become properly communal and in touch with the Earth. It is to these people that this book is particularly dedicated. They have been my inspiration and my challenge, as I have joined in teams with others to do things like found our own local currency, known as the Bristol Pound, develop a community-supported agriculture project on a shared-risk, shared-harvest basis, open up the very vicarage I used to live in as a community garden project and, most recently, launch a food and enterprise co-operative known as Real Economy.

It is these experiences that form the basis for the reflections about good practice in the last section of the book. There are some key ideas in these chapters about how we can creatively harness our passions and interests and incorporate new approaches to the sacred in the service of the planet. At the same time there are vital watchwords about avoiding burn-out and creating a strong inner life.

This book is written for those who want to achieve social change. Some are existing activists, who I hope will find encouragement, confidence and clear ways of thinking that will undergird their search for a different kind of human society. Some are waiting in the wings of the theatre of change, aware that something new is coming to the world, but not clear what they can do to help bring it to birth. Sometimes they feel overwhelmed, sometimes guilty. They long to do something, but cannot quite see what or how. I hope these people will find here something like a manual for social change that will help them get stuck in with all the passion and energy that is beating within.

The book begins with a restatement of some human capacities and how our present culture has fixed on certain human abilities, but has ignored or denigrated others. Building on the magisterial work by Ian McGilchrist, *The Master and His Emissary: the divided brain and the making of the Western world*, it suggests that what we call 'Western' society is now seriously lacking in several areas, including our capacity to communicate in non-verbal ways, our ability to create cohesive communities, our ability to imagine the feelings of others , our recognition of the commitments that predetermine our decisions and our ability to perceive life as a whole, including the interconnectedness of all things. I draw on the extensive work of Frans de Waal on primate and other animal societies to remind us of some of these capacities and suggest that non-verbal communication is a route to a renewed sense of the 'spiritual' as we have come to understand it. It becomes clear that the human species is nothing like as rational as it pretends to be. We have to approach the world using pre-formed commitments in order to deal with the complexity of life and we treat these commitments with a profound defensiveness. We hate to change them. I review Jonathan Haidt's seminal work in this area, particularly the importance of pre-formed commitments in the world of politics.

In the central section of the book, I outline how science, politics, economics and religion are all failing to respond to the challenges of this moment in human history. It is possible to define immediate causes of these institutional failures and I review the findings of some of the most astute commentators of our time in each of these fields. Yet understanding the logic of institutional failure does not solve the problem, because something else is going on here. The reason for our institutional failures is not simply that science has been undermined by sceptics, who are fueled by a media hunger for controversy, nor that economics is intrinsically short-sighted, nor that religion is human-centered; the deeper reason for each of these institutional failures is the inner weakness of the general populace. In each case I attempt to define what particular weakness is exacerbating the failure of our great institutions. It may be that democracy itself is both a great strength of our societies and a pernicious weakness. As all institutions become more responsive to the people then the institutions can find themselves locked in a tragic embrace with the people's own inner failures, responding reflexively to the latest trend in language and behavior and unable to defend the people against their own worst excesses. It is like society becomes a body without a backbone.

The last section of the book begins a process of addressing the issues that have been revealed. It is clear that we need a new strategy for our common life. The established ways of coordinating human society are tired and ineffective. Only a new strategy can give us hope and overcome the sense of being overwhelmed by the environmental challenges with which we are faced. So this section argues for a shift in perspective from the nation as the focus for change, to the city region. It proposes that we need to localize those parts of the economy that can be localized by taking concern for the cities of the world and their hinterland. Such a process of localization will not only create local wealth, but also make for strong and diverse expressions of community within and around our cities. Sustainable city regions may provide us with the new focus we need to imagine global sustainability.

It also considers our inner lives, which are the source of our energy to work with others, yet are so often dogged by internal contradictions that weigh us down. We are called to acknowledge 'The Fire Within' our lives, those root beliefs, values and commitments that we hold so important, then learn to understand and work creatively with this sacred center within us.

The renewal of a sense of working together is a subject of vital importance in our individualistic, pleasure-seeking world. There is a real possibility that the world could unite around the challenges that we are faced with today, but to do so will require a new approach, which recognizes the values we share, and engages in new enterprises that together change the fabric of society. Imagination, empathy, dedication, community and celebration are the hallmarks of this new approach. Examples are given of how this can work out in practice, including the use of music on a global scale and the consecration of land as sacred in a new sense. People are encouraged not to just turn the handle in paid but meaningless work, but to take the risk of stepping out and embracing 'meaningful work', co-operating with others around 'Shared Values' and 'Clear Purpose' with the overall aim of social transformation.

In the end, it is up to us. We can shuffle off quietly to our own demise or we can rise up as individuals and communities to reshape the world. The last chapter is a call to creativity and purpose drawing together the threads of the book into an appeal to live life to the fullest.

February 2016

Part I

Contours

High in the sky
the eagle soared
with watchful eye
on the hills and valleys
rivers and streams

Chapter One

Life beyond Words

In a Nutshell

In this opening section of the book, I want to consider how we perceive the world around us. I am going to suggest that there are vast realms of human experience that we currently do not notice, or tragically undervalue, because of the culture that dominates the Western world. Bringing our whole set of perceptions into balance may give us a much better grasp on reality. One way to open up this subject is to consider those elements of our lives that are 'beyond words'; that is, existing prior to language, because these remain of vital importance to us.

The Workings of the Mind

My thinking in this area has been profoundly shaped by the work of Ian McGilchrist on the structure of the brain. His book *The Master and His Emissary: the divided brain and the making of the Western world*[1] reveals some extraordinary features of the human brain and its impact on culture. He has no time for the populist and simplistic statements that are so prevalent in our media about left- and right-brain activities, and yet he is convinced that more careful thinking about the structure of the brain can give us important insights into our behavior and our culture. The brain, it seems, should not be considered as a static structure, but in terms of its complex processes and flows, or 'ways of working' that it can adopt. It is also clear that the workings of the mind and the cultures of human society affect one another such that changes in one can reinforce changes in the other.[2] It seems that Western Society is currently working with a very odd and inadequate set of mind processes.

The story goes something like this. The brain is structurally divided into a right and left hemisphere joined at their base by the corpus callosum which mediates relations between the two sides. Most brain processes involve activity in both parts of the brain, but the importance of each hemisphere to a particular process is not necessarily equal and can be revealed by a combination of imaging techniques and studies of the brain in people with particular lesions due to trauma or stroke etc. It seems that the left hemisphere is particularly concerned with *representing* the world in terms of signs and symbols and their interrelationship. It categorizes and classifies. It provides 'focused attention' on detail and constructs a static view of the world from a sum of its parts.

By contrast the right hemisphere is about *experiencing* the world. It perceives the world in relationships, connecting disparate ideas and emotions, working with metaphor and holding big-picture views of the world. It seems that all new ideas originate in the right hemisphere and it is the source of creativity.

The evolutionary origins of this distinction between the hemispheres can be seen in the bird feeding on the lawn. Have you noticed how it tips its head so as to turn its right eye to the ground and its left eye to the sky? The right eye connects to the left hemisphere and gives the bird 'focused' attention on the ground so that it can spot its food, while the left eye, connecting to the right hemisphere, gives it a very different 'vigilant' attention all around watching for predators. Vigilant attention is an open process watching everything for what might be unusual, right into the extremities of vision.

Careful analysis of brain processes leads to the following understanding of the roles of the two hemispheres in terms of our perception of the world.

Left hemisphere-dominated perception:
creating a representation of reality
tendency to abstract things from context
using signs and symbols
categorizing people and things
sees science as a mechanism
science creates a manufactured world of unconnected objects

Right hemisphere-dominated perception:
experiencing the world as a totality in context
connectedness
including relationships and emotions
working with big-picture ideas
using metaphor, poetry, creativity, humor
nature presents reality to us as an unmanufactured whole

So it is that we have two quite different ways of perceiving the world available to us as humans. They are both important to a properly functioning society but they are not *equally* important. It transpires from studies of people with particular brain lesions that the left hemisphere serves the right hemisphere, but does not do well without it. So a person with a stroke in the right hemisphere 'perceives' the world through their left hemisphere only. Their eyesight might be OK, but they may only recognize the right-hand side of the field of view before them. So, clocks appear without a left side. They can even consider their body to be missing its left arm or leg. By contrast when there is a left-hemisphere lesion the right hemisphere perceives the whole field of view, albeit lacking some detail, due to the absence of focused attention.

These very strange observations have important repercussions in terms of the way our minds are working today. It is apparent that our focus on science and technology, and our view of the world as a machine, is a left brain-dominated process and produces a very skewed society. As a result we see the world as a mass of constituent objects that we must categorize and reason from. People are viewed as 'categories' not as unique individuals and society becomes formulaic. After a fascinating review of human culture through the ages and the impact of the left and right hemispheres of the brain, Ian McGilchrist sums up the left brain-dominated society in phrases like:

> *...loss of big picture, focus on detail, specialized pursuit of knowledge, information gathering, failure to value skills or experience, importance of paper qualifications, bureaucratic domination of workplace, control, quantity replaces quality as goal, people defined by categories, morality/values measured only by utility, talk of liberty but increased surveillance, lack of will-power/self-motivation except for greed and desire to manipulate, people are passive, religion as mere fantasy, power and control through 'factual' worldview, spectators rather than actors, history not important.*[3]

I don't know if you recognize our society in that description? I certainly do. It suggests that many of our issues today have to do with a lack of balance in our minds and in our culture. Why do we hear the science about climate change, but not *feel* its implications? Maybe it is because we have lost the capacity to feel these things. We no longer feel for our relationship with the Earth. Why do we feel so unhappy and unfulfilled? Perhaps it is because we are failing to provide a proper focus on developing a rich set of relationships between people? Why are we so obsessed with our bodies? Perhaps it is because we see them as things or worry about our self-image? Why do we feel an antagonism between science and religion? Maybe the heart of it is that there are two quite different ways through which human beings can pay attention to the world? We need to correct this balance if we are to function well as a society. These themes will echo through the chapters that follow in this book as I try to work out their implications for a reform of our lives and our society.

Life beyond Words

I was standing behind a screen. I slipped a very uncomfortable red blob onto my nose and stepped out. Our task was simple. It was to walk across the makeshift stage to the safety of a screen on the other side, while turning just once toward the eager audience. "It can't be that hard," I thought. I was used to public speaking after all. But here you did not speak. You just looked. And they looked back. And something happened.

I was on a clowning course led by my friend Angela Knowles. As the weekend progressed so we engaged in more and more sustained acts of 'clowning'. I rapidly found that if I tried to be funny, it appeared hopelessly contrived, but sometimes I would do something, almost by mistake, that the audience found interesting. I would play on that theme by doing it again, or doing it differently, and soon the people were in fits of uncontrollable laughter. I was learning to respond to people without using words.

It is easy to imagine that clowning is a light thing, a superficial pursuit to distract us from real life, but Angela did these courses because she was convinced that clowning somehow connected with spirituality. By getting in touch with life beyond words we could reach into forms of self-awareness and expression that are simply not available to us in our modern culture. That evening she told us of a man who visited a psychiatrist in Paris with a deep depression that had dogged him for years. The psychiatrist was at the end of her tether. What could be done to help this man, now that all the conventional therapies had failed? The psychiatrist said, as a last resort, "There is a clown in Paris. He is one of the most famous clowns of all time. Why don't you go and see him perform?" The man looked at the psychiatrist with his deep doleful eyes, and he said, "I am that clown."

That little story points us to a deep truth about clowning. Sometimes the clown is a grief bearer. We laugh at the clown's misfortune. We are captivated by their sorrow. I once saw this expressed in theater. Pete Postlethwaite was taking the part of *Scaramouche Jones,* a one-act play written by Justin Butcher that takes us through the life, and the tragedy, of a clown as the events of human history cross his path. It was a tour de force of acting. Pete Postlethwaite held the audience for an hour and a half, just him, talking through the life of this clown. At the end the audience rose to their feet and roared their approval. The play was, of course, expressed in words, but it was so much more than words. Great actors pour themselves into their characters, so that every nuance of their body, their movements, their eyes and their actions express the role they are inhabiting. Actors know about life beyond words. And they know about spirituality.

Laurence Anthony lived in South Africa and came to hear about a rogue herd of elephants that were in danger. They were behaving badly and considered a danger to humans. Some were calling for them to be shot. A terrible deadline was set. If Laurence could not receive them within two weeks, then they would be disposed of. Laurence had no experience of elephants, but he was a remarkable man. He rapidly built a 'boma', a small compound, to hold them within the Khula Khula reserve, which he managed. Then he sent for the elephants. By the time they arrived, they were seriously disturbed. The original matriarch and her young calf had been shot and killed during the capture. The very first night the elephants broke out of the compound by deliberately felling a tree so as to collapse and disable the electrified fence. Local people went out looking for them. They had guns, because now the elephants were considered 'fair game'. After a traumatic chase Laurence Anthony managed to drive them back to the boma with the help of a friend with a helicopter. It was then that he realized that his only chance of saving this herd would be to get deeply in touch with these animals. One more breakout would be the end of them. He camped just outside the fence waiting to see what would happen. Every morning the new matriarch, known as Nana, would approach him at the fence, ears flapping in an aggressive pose as if she was about to break out and attack him. He knew that, if she decided to, she could just walk through the flimsy fence. So he tried to soothe her, speaking to her in gentle tones, using words he knew she did not understand, but hoping that she would pick up on his intention to help her and her herd. Each morning this stand-off at the fence repeated itself. Laurence would back off as she approached, showing deference, but staying relatively close. Finally, there came a morning when Nana approached without the usual aggressive body language. Laurence noticed, and retreated less than usual. Then, very gently, the elephant reached her trunk over the fence. Laurence gingerly put forward his hand. And they touched. It was a moment of obvious trust, evidenced by the trail of pachyderm slime that Laurence found on his shirt afterwards! The elephants were released into the nature reserve shortly afterwards and began an extraordinary relationship with this man, who said, "I did not adopt the elephants, they adopted me." The full story is told in his fascinating book *The Elephant Whisperer.*[4]

Part of the wonder of elephants is their extraordinary abilities of communication. Our phrase 'a rumble in the jungle' relates to the deep sounds that elephants use to communicate among the herd. A rumble can carry through deep forest for 2.5kms or even further in particular seasons and places. It seems that the rumbles allow them to clock each other's presence, and communicate things about their situation. Elephants also seem to have a special ability to sense when rains have come, even hundreds of miles away. Setting out on a migration in such inhospitable terrain spells certain death unless there is water at the end of the journey.

How they know about the rain is a mystery, but human tribes in Africa have learnt to set out when the elephants do. They trust that the elephants know when they should go.

Laurence Anthony's life came to an end unexpectedly when he died of a heart attack at age 61. At the time of his death, there had been no elephants around his home for a year and a half. They were spread across the vast nature reserve. Yet, when he died, they came. How could these great animals know that this human, whom they had adopted, had died? Within 48 hours of his death, two herds visited his home. They stayed for two days, as if in memory of him and, according to his wife, they have returned each year on the anniversary of his death.

Elephants may be showing us something of the amazing world of communication that is known by social animals. Some of these things we understand and some we don't. Animals clearly have abilities we don't have. They may also show us some abilities that we have lost through our word-dominated culture. Think of creatures like the whale, the dolphin, the meerkat, the wolf or even our domestic dog. They have phenomenally complex and successful social systems, all of which are mediated without words. Could it be that we human beings have allowed words to supplant some of our natural abilities in this area? Do we simply fail to appreciate the importance of the touch, the glance and the smile? Have we lost the ability to read the concerns and interests of others from their demeanor? These are important questions and they may be a clue to that other failing of modern Western culture, namely the loss of any meaningful spirituality.

I was recently on holiday in Fowey, a delightful port in Cornwall. We had crossed on the pedestrian ferry from Polruan and sat down with some others by a coffee shop looking out onto the Fowey estuary. There were, perhaps, a dozen other couples doing the same thing. And I wondered just what it was that was going on? People were staring at the view. There were snippets of conversation, but somehow it was more than that. We were sat there experiencing life. We were there in a whole sense, hearing, seeing and feeling together with the whole complex of emotions that moved within us like the waves of the sea that lay before us.

Then it occurred to me that this is *most* of what we do in life. We are in it, immersed in it, as whole beings and we experience it as a whole. Most of our conversations are not really 'about' much. They are superficial, merely encouraging that great flow of experience within us and between us. This explains why the higher social animals can do so well without language. We may still have all those capabilities, but have grafted on language in such a way that we tend to forget that great undertow of subconscious communication that is going on within us and around us. As I walk around the park with my dog, I make casual remarks to strangers about the weather, or some local happening, while my dog greets another dog with a mutual sniff of the backside. Both are really the same thing. A hallo, how are you, don't be afraid, everything is okay. In one sense they are just a ripple on the surface of our lives, but they are also an enrichment, giving a sense of place and belonging.

Music and Art

It is clear that some animals have a facility for music. Birds, whales, dolphins, elephants and even mice use pitch and rhythm as a means of communication. Elephant vocalizations actually span four octaves even though we cannot always hear them. It seems that music existed prior to language in evolution and grew among some social animals as a means of managing their society. Music has a similar, but less complex, structure than language, as shown by the fact that children have the capacity to sing before they can speak.[5] More importantly for the collective, music conveys emotions. Humans are well aware of the emotional stirrings that music provides, but we may not have recognized how important this is in cultural formation. McGilchrist suggests that music was originally passed on between people in a process of mimicry. It was not taught, as we have come to assume, by applying pupils to a series of written notes on a page, but by people hearing music and trying to repeat it, or respond to it. The joy of the choir, which many still experience today, is the joy of a collective act of mimicry. In traditional societies the musical celebrations, dirges and dances were part of what *made* the community. McGilchrist says that music "is a vital way of binding people together, helping them to be aware of their shared humanity, shared feelings and experiences, and actively drawing them together."[5] He speculates that music may have taken over as the primary means of bonding between people when the community became too large for physical acts, like grooming, to be affective. An interesting animal analogy for this process is the Siamong Gibbon, where male/female pairing is accompanied not only by mutual grooming but by 'duetting', where the normal territorial call is enhanced responsively as the pair announce the strength of their relationship or, we might say, celebrate their love.[6]

It is worth noting here that communal dancing, which is, of course, related to music, is an important feature of the life of every society.[8] Why should every society find music and dance important? In this vital realm of life beyond words, actions as well as rhythm and tune can unite us. To move in time with music is to allow the music to unite us. It is for that one glorious moment to let go our individuality and relax into being a community. The joy of such unity can be overwhelming.

It is hard to write about art. It needs to be experienced rather than analyzed, but it must certainly feature in any discussion of this realm. Turner, one of our greatest UK artists, was a man of few words, who poured himself into his painting. He found himself painting in a style and fluency like no one else around him at the time and he was occasionally mocked by the public.

Within him, it seems, was an enormous emotional/spiritual drive to express himself through art. It is a drive that can be recognized in others, who devote their lives to the arts as a calling. Michelangelo's life, for example, is a profound example of dedication to sculpture. Such people feel the world and see the world in a way that they know to be beyond words. Their works are to be experienced, not talked about, not reduced to components – but experienced as a whole. They are not about simple material representation. That is the domain of science. They present us with their whole experience, including their inner emotions and passions. For example, as Turner bears down on an old and revered sailing ship being towed to its final destruction by a black, steam-powered tug, it evokes his most famous *Fighting Temeraire* painting with a sense of confusion and depth and with a light that seems to trigger an emotional perception of what is going on as old worlds fade and new, suspicious yet powerful realities dawn upon us. Such is the stuff of art.

The application of artistic sensitivities is particularly obvious in the architect, who tries to imagine not only the shape and function of buildings and spaces, but how such a building or community will feel when people live or work there. This process, at best, engages with the right-brained, relational and imaginative side of our thinking and feeling and is very much beyond simple verbal definition. Many of us know bleak and dull spaces that have been designed by architects, as well as others that are vibrant and full of hope. Some areas seem unable to escape from their tragic history without wholesale architectural redevelopment. What is it then about architecture and architects that is so important? Perhaps it is that a good architect can *feel* the outcome of their work. They can translate from drawings on a page and consultations with a community, to a powerful practical imagination of lives being lived in all their complexity.

In Bristol, UK, where I live, we appointed our first ever elected mayor in 2012. We chose a person who was independent of party politics and was an architect of many years' experience. I mention it because his election caused a certain frisson of expectation, even excitement, in the city. I have thought about why this happened and I think that some of the attraction of this man was that he was not your regular political type. He was not bowed down by years of political infighting. He did not spout the usual party lines and yet was clearly aware of realities of power. And I guess some of our interest came from the fact that he was an imaginative person, interested in design, who was not afraid of fun and creativity and who could make new things happen in the city. I suspect that it was partly because of his leadership that we were awarded the accolade of Green Capital of Europe for 2015. Creative people can change things. We need to recognize that much that is of value to us as human beings exists in the realm of life beyond words. And people who know that are important to our future.

The contrast to good design may be found in the modern GP surgery. As you walk into my surgery there are no people to greet you. A computer screen on the left wall invites you to enter your date of birth, then tells you to sit down and wait for your appointment. Chairs are arranged to face a screen, which carries warnings about things, tells you not to behave badly, and occasionally pings and a notice comes up: "So-and-so to the see Dr… in room number 5." Such designs are based on 'efficiency' as measured by staff time and money. Someone has taken the view that having a person in the room to welcome patients is not the best use of money. And yet, the warmth of a human greeting is surely a vital component of healing? Much of that greeting is expressed in body language, the smile, the reassuring tone that conveys "it's alright now, you will be seen, we can help" etc. The patients express their discomfort about my surgery by not recommending it to their friends. It is rated as one of the worst (25%) in the country for referrals from friends.

A Relationship with the Natural World

I now live on a modern suburban estate in North-East Bristol. The houses are conventional modern constructions, with a mix of facades, some apparently grander than others, but all very much what you would expect. The houses are placed along a meandering set of roads with speed bumps, designed to discourage rat running and maintain a tranquil space. Yet, to the credit of the designers, the development includes a set of original country paths with all their diverse hedging, all protected as footpaths and bridleways and providing routes into the experience of the natural world. Within fifty yards of my house, I can enter a shaded path that has been almost untouched for a century or more. It is rich in wildlife, and a truly beautiful space to be in.

Beyond that is the park, a substantial area of perhaps five acres with some ancient trees and another set of trees that have grown since about 1997. There is the 'tidy' side of the park with two enclosed areas for children and much grass, and a wilder side, which has been much more interesting, to my mind. This wild side of the park was not 'managed' for around fifteen years and went through a whole set of stages of growth heading for the woodland, which is the natural end to unmanaged landscapes in England. By 2013, it had become a thick mass of saplings, where ash, oak and hazel all abounded. Children played in the undergrowth. Dog walkers, like myself, and many others marked the seasons as they waited for sloes and blackberries, listened to the birds, and experienced a deep joy simply through being in the space. Yet in 2014 there were some traumatic changes. The local council suddenly decided to 'manage' the space. Citing the need to maintain flood defenses, in relation to streams running through the park, they removed almost all the re-growth and the park was rendered bare. Just a few oak saplings remained. There were some, of course, who preferred the new open vistas and felt more confident in a tightly managed public space, but for others, the changes were devastating. One friend said to me, "It has been butchered" and declared that he was avoiding walking in certain areas because of the grief it caused him. Another commented that, for them, the space was sort of "sacred" and the changes had felt like a desecration, a despoiling of something holy.

These feelings are a clue to something important, which I invite the reader to reflect on in this opening chapter. If we immerse ourselves in the natural world over a sustained period of time, we experience something. We gradually come to appreciate the enormous richness of the life around us. We are surprised by new insights and discoveries that dawn on us unexpectedly and, at the same time, we are reassured by the rhythms of life through the seasons. We sense that we are part of something bigger than ourselves. Something deep within us is settled by this experience. We are part of this world and it is part of us. I sometimes reflect on springtime and think to myself "How many more times will I experience this delight?" The answer is "I don't know." Maybe many. Maybe few. So it is with much of the life around me. We live today, but we do not know about tomorrow. And we live today in the company of the 'community of creation'[9] to which we are irrevocably joined.

The poet John Clare was, at one level, a simple soul. His language carried strong echoes of his local dialect, but his heart was full of the natural world and he wrote some of the most evocative and passionate poetry about nature that has ever been written. Poetry is, of course, ultimately a verbal means of communication, but the formation of this poetry was only possible because John Clare first of all *experienced* nature and it is that experience of which his poetry tells. This is what Clare tells of himself:

I grew so much into the quiet love of nature's preserves that I was never easy but when I was in the fields passing my sabbaths and

> *leisure with the shepherds and herdboys as fancys prompted sometimes playing at marbles on the smooth-beaten sheeptracks or leapfrog among the thymy molehills sometimes running among the corn to get the red & blue flowers for cockades to play at soldiers or running into the woods to hunt strawberries or stealing peas in churchtime when the owners were safe to boil at the gypseys fire who went half-shares at our stolen luxury we heard the bells chime but the field was our church.*[10]

People come to a similar experience of the natural world in a variety of ways. They may walk up mountains, sail on the sea, grow their food, or simply tend a small garden plot, but all these people would share in the transcendent experience of the natural world. It does something to them. It takes them out of themselves. It provides a deep assurance and profound joy. People who know this are, in my understanding, getting in touch with something beyond our simple material existence, which I would call the realm of the sacred.

Why Is This Chapter Important?

In this chapter I have tried to introduce us to a range of experience that we have access to as social animals and that is not simply defined or adequately conveyed by means of words. This realm exists prior to words. It is known to all social animals with complex minds. I wonder whether the key to accessing true spiritual experience is also about gaining access to this state prior to language. Could it be that trying to put everything into words has actually limited our experience of life?

One person who has explored such thinking is Nicholas Carr.[11] His book, *The Shallows,* is focused particularly on what using the internet is doing to our minds, but his more general thesis is that every piece of technology that we embrace changes the way our mind works. So, for example, when human beings developed maps, we were able to chart the world, but at the same time, some of our natural abilities to remember territory, find our way home etc. were lessened. We became dependent on the map to guide us. Similarly the invention of the clock gave us greater precision in time, but led to the loss of a more primitive sense of time based on the progression of the sun in the sky. We no longer *felt* the time as we had used to, but relied on our watches. He also suggests that the development of writing might have given us a more precise record of culture, but led to a loss of community memory, which had been so strong in oral cultures. Nicholas Carr concludes that our embrace of internet technology is far from an unalloyed good. He proposes that much internet use is characterized by superficial interaction with immediate feedback. It thereby occupies the processing parts of our brain very actively, but our ability to sustain organized thought or engage in deep reflection may be seriously diminishing.

It is also obvious from this that the so-called 'social media' are lacking something important. They are shorn of almost all the components of non-verbal communication that social animals have evolved to use. This is why it is so easy for people to project false personas on the web. There are a host of non-verbal signs of integrity that we look for when someone is telling us something on a face-to-face level, but on the internet we can pretend to things easily. We also confuse public and private on the internet for similar reasons.

It is clear from the work of Ian McGilchrist that Western culture is now working with a very inadequate perception of the world. The workings of our minds have become skewed so as to no longer properly value the holistic, emotional and dynamic processes that are the creative source of our lives. Engaging with life beyond words is the first step to acknowledging this and building a renewed appreciation of life in all its richness.

Chapter Two

Feeling Our Way

In a Nutshell

In this chapter I am going to suggest that we no longer recognize the importance of feelings in our public life. This is not to say that we have become stoics or something, whereby we suppress feelings, but it is rather that we no longer acknowledge the importance of feelings in shaping our societies. The attempt to found society on science and reason alone has been doggedly pursued by academics and policymakers now for several hundred years, but it has led us down a very strange path whereby feelings and other associated deep-level motivations are seen as trivial compared to ideas. I will argue in this chapter that our emotions are still the most important factor in shaping human community life and need to be recognized as such.

Feelings in Public Life

Robin Dunbar in his book *Primate Social Systems*[1] tells a story of Barbary Macaque monkeys on Gibraltar. It seems that when trouble is brewing between two males, one of them frequently reaches out for an infant and places it between itself and the other quarrelling male. The argument between the monkeys then typically subsides. What is going on here? Every ordinary person who hears this story understands immediately that the young one evokes compassionate feelings that diffuse the tension. Yet, if you asked an academic about this thirty years ago, when the book was written, they would have been much more circumspect. Their scientific minds hesitated to even talk about the inner workings of the minds of animals. They deliberately confined their thinking to evidence of actual behavior and any genetic advantage that might result to the animal. In consequence they were hard-pushed to come up with any understanding about this Macaque behavior. The scientific method simply could not allow them to talk about the very uncertain and dynamic world of animals' feelings. As a result they were unable to understand what was going on.

This approach was popularized by Richard Dawkins in *The Selfish Gene*[2] and described the animal as mere housing for a gene, whose only organizing function was self-replication. It gave impressive clarity and apparent rigor to the scientific study of animals, but strictly limited what a scientist might say about the mind of an animal or their community life. It also determined that the baseline for all animal-behavior studies was to be the individual, even for social animals, because this was the 'unit' of genetic replication. Such thinking meant that community life itself became an enigma, a problem to be solved, because you had to show how community life and all its various manifestations were actually advantageous to the individual. Much ink was spilt over the 'problems' of altruism or the evolution of trust, for example.[3] And one of the terrible failures of this whole approach was that scientists refused to talk about the inner world of animals with an appropriate seriousness.[4]

One man who has stood against the tide in this respect is Frans de Waal. He is a lifelong primate psychologist who has dared to study primate behavior as a means of revealing the dynamics of their inner lives. I say 'dared' because he has taken considerable stick from academic colleagues about his methodology for decades, yet the value of his work is now, at last, widely recognized. His research has included long-term studies of primate communities, demonstrating clear examples of hierarchical behavior and alliances that are very similar to those we recognize in human societies. He has also focused attention on altruistic and emotional behavior. He tells this story, for example, about the last days of an alpha-male chimpanzee known as Amos, who was dying of cancer:

> *Amos didn't betray his condition until the day before his death when we found him panting at the rate of sixty breaths per minute, with sweat pouring from his face, sitting on a burlap sack in the night cages while the other chimps were outside in the sun. The other chimps kept returning indoors to check on him, so we cracked open the door behind which Amos sat to permit contact. Amos placed himself next to the opening and a female, Daisy, gently took his head to groom the soft spot behind his ears. Then she started pushing large amounts of excelsior through the crack. This is a wood shaving that chimps love to build nests with. They arrange it all around them and sleep on it. After Daisy had given Amos the wood wool, we saw a male do the same. Since Amos was sitting with his back against the wall and not doing much with the excelsior, Daisy reached in several times to stuff it between his back and the wall. This was remarkable. Didn't it suggest that Daisy realised that Amos must be uncomfortable and he would be better off leaning against something soft, similar to the way we arrange pillows behind a patient in the hospital.*[5]

You can see how this story is told. It is based on evidence. This is what happened, yet it moves on to make an inference about the primate's inner life and affections with a view to demonstrating the richness of the emotional life of primates. Here is great empathy, even love, expressed in the actions of the young chimp.

Many social animals share a rich emotional life. Frans de Waal proposes, and many now concur with, the idea that the world of our feelings actually derives in an evolutionary sense from the care of the young.[6] Animals with complex minds that have few offspring and provide them with sustained care tend to develop social systems. And the reason is clear. The parent learns to imagine the feelings of another. They develop empathy. From this root of empathic concern comes the whole realm of emotional relationships in a society of social animals. Particular emotions are, of course, all felt to different degrees by different people, but these 'passions' give rise to great joy and heartache among us and are the very heart of our lives. They are important not just to individuals, but also to the society in which they participate, because they become the root of our deepest concerns and values. Chapter Four will describe how such personal differences of feeling translate into contested stories that lie at the heart of the listening and negotiation on which a good society depends.

Empathy is a particularly important ability of the social animal. The ability to imagine how others feel allows us to reach out beyond our immediate family and social circle to embrace others in the community, and even strangers. The great question about human societies tends to turn around how it treats those who appear different from themselves. The campaigners for the abolition of the slave trade used the phrases, "Am I am man", "Am I a brother" etc. as an appeal to people. It was an appeal for empathy. Similarly in the UK today our view of migrants might depend on our empathic reach. There was an extraordinary moment in the summer of 2015, as the 'migrant' crisis began to be felt right across Europe. People were fleeing from Syria, Afghanistan and Iraq in enormous numbers. Our Prime Minister used threat language, talking about 'swarms' of people trying to cross the channel from Calais, and announced new funds to build fences. And then there was a young boy washed up on the beach who had set sail from the Southern Mediterranean coast in a flimsy craft. Someone filmed the moment that his lifeless body was scooped up into the arms of an aid worker. That clip went round the world and the whole tone of public debate changed. Something had to be done. Communities readied themselves to receive these people, now called by their proper title, 'refugees'. Even our Prime Minister felt obliged to take a more compassionate approach. The development of empathy might be one of the best measures of progress in any social community.

The Range of our Motivations and Desires

Social systems in animals with complex minds[7] appear to feature similar sets of motivations and desires that are formed in relationship with their social situations. We would not all describe them in the same way, but I list some of them here as three primary motivational sets: the loves, the obligations and the anxieties about power. Each set has a list of associated feelings that are generally felt to be less honorable or desirable, such as grief, loneliness and jealousy, which arise when one or more of the desires is not satisfied.

The loves:
Compassion/empathy
Friendship
For soul friend
For young
For weak, marginalized
For community, people, nation
For the natural world
For your children
Sexual love
Lust
And their associations:
Grief
Lovesickness
Jealousy
Loneliness

The obligations:
Return of favors

Loyalty
Justice
Forgiveness
Grace/altruism
Mercy
Kindness
And their associations:
Cheating
Enmity
Revenge
Aggression

Power anxieties:
Ambition
Pride
Desire to be 'in'
Holding territory
Discipline
Desire for order
Recognition
Desire for creative freedom
And their associations:
Anger
Envy
Scapegoating
Humiliation

Such feelings are surely the warp and weft of any society and how we work with them creates the distinctiveness of different communities.

The writings of Frans de Waal give powerful evidence of the complexity of the societies formed by social animals, but they also raise a question. If these animals do not use words, how do they structure their social lives? The answer must surely be that their societies are constructed through a non-verbal negotiation based on their feelings and associated motivations as illustrated here. As word-based animals we tend to shy away from acknowledging how important such feelings are to our own human societies. We like to believe that we negotiate how to live in society through words and the articulation of ideas, but in practice we also rely substantially on feelings, like the other social animals. If we can recover a proper awareness of this, then we might build our communities better in the future.

One way to get in touch with this area is to consider the emotions of love and grief. Grief is often a sign of love and it is felt deeply by many higher animals. Barbara King[8] has chronicled stories of cats, dogs, horses and elephants and even inter-species animal friendships that demonstrate signs of grief when a companion animal dies. Typically the animal seeks out the one who is lost, may visit the dead body or the grave, will show signs of anxiety, is likely to go off food and, in extreme cases, may even die. Barbara King tells the story of a horse whose leg was irreparably damaged and had to be put down, but who was buried in the field in which she had run with the other horses. The human carer of the horse visited the field on the evening after the burial to lay some flowers on the grave, and as she stood there in the growing darkness, she became aware of horses gathering. Six horses gathered with her at the grave, all those horses that had been close to the deceased, and they just stood there in silence, with their heads down, looking at the grave. The horses stood there all night, in their vigil.

A neighbor of mine died recently. He was of a Romany family and the son came to us and told us that the family would be gathering the night before the funeral. Cecil, the man who had died, would be 'coming home' for the night and there would be a fire. So it was. A fire was lit. A tent was raised in the garden and the family gathered around the deceased and stayed there all night. "It is our way," said the son.

Stories of power and corruption are also vehicles for thinking about our inner life. Human societies are replete with examples of hierarchy and status-seeking, as well as some inspiring examples of more just and empathetic behavior. It may come as a shock to realize that many primates illustrate very similar behaviors. Their hierarches have strong parallels with our era of monarchy, their disputes are similar to our rebellions and their peacemaking strategies are so subtle and interesting I think we might learn from them.[9]

The Evolution of Institutions

Emotions may be the most powerful instruments of human relationship we have available. Consider, for a moment, the family. For some, it is a wonderful institution; for others, a place of torment. And the difference? I would suggest that successful families have learnt to appreciate and work with the emotions of their members. Nothing is more conducive to a sense of well-being than to be in a community where there is mutual emotional understanding. On the other hand the most debilitating periods of our lives, whether in families, workplaces or elsewhere, tend to be when there are unresolved emotional issues. So-and-so is cross with someone else, holding a grudge, abusing their power. There are many possibilities, but the results are the same. The members of such communities are like walking wounded, hurt people, who can scarcely function.

Consider the 'atmosphere', as we call it, when we enter a room where there has been an unresolved emotional engagement. Maybe we can sense it in the body language. The slightly frightened looks in the eye, the awkward posture or stilted phrases. But there is no doubt about its presence.

I mention these things as a way into larger questions, because these things are part of our everyday experience. I am going to propose that our primary human institutions arise from such emotional engagements. Let me begin with a story.

Frans de Waal was working with some capuchin monkeys and giving treats for rewards at certain tasks. Capuchins apparently rate their rewards so that cucumber is okay, while grape is great. Anyway, one capuchin was in a carefully constructed enclosure where he could watch other monkeys doing things, but not interfere with them. The capuchin would do a task and receive a cucumber in reward. He did this contentedly through several cycles of task/reward, then he saw something. A neighboring monkey was doing exactly the same task and being given a grape. The original capuchin went mad with fury.

A film of this event was then shown to a human audience, who burst out laughing at the tantrum thrown by the wronged capuchin. These people immediately recognized our own struggles for just reward as measured against the reward of our peers, like a dispute about a pay scale.

In a further elaboration of this experiment de Waal was able to show another development. Sometimes the subject who received the higher reward looked uncomfortable. It could not be seen to have a better reward than the others. The monkey would then be seen to signal in such a way as to say to the researcher that the other should have a similar reward to itself.[10]

These things are not trivial. What is being revealed here is a basic sense of justice, something that is powerfully formative of all economic systems. It suggests, for example, that currency may be a device which, at heart, is *about* fair exchange. Grapes and cucumbers are not of equivalent value to a capuchin. We can imagine in early human societies that the demand for fair reward might easily have led to the development of a currency which was of settled value and which could be exchanged in an open and accountable way.

Since the Enlightenment and the work of Adam Smith and John Stuart Mill, we tend to think of economics as a theory, based on formulae, which are built around human self-interest. Yet the self-interested human being of these theories is nothing but a hopeless caricature of the complex people we are. In fact humans bring a whole range of emotions to bear on their economies and these must be appreciated if we are to live well. The sense of fair play alluded to here, especially in the second example, clearly derives from an empathic imagination. We can put ourselves in another's shoes and feel uncomfortable if fair treatment is not meted out.

Such, I would argue, is the true ground of economic systems. They are about fair exchange. If a fair, accountable system of exchange can be developed, then individuals can commit themselves to specialized trades and manufactures assured that they will get a good price at market and so can provide for the rest of their needs. In this way a good economic system makes space for economic creativity, diversity and technological invention.

As an aside it is worth pointing out that while many of us may be blind to the emotional roots of economic systems, there is one group of people who are well aware of the power of feelings in this area and work to exploit them mercilessly. They are the advertisers. Since the early part of the twentieth century the advertising world has recognized the power of associating positive feelings with a product. It was Edward Bernays, a cousin of Sigmund Freud, who first made the connection in the 1920s.[11] For him this was a path to a sort of peace, a way to keep people happy when there was no war to occupy them. You got people to buy things and made them feel good while they did it. Advertisers would achieve this by associating feelings that people wanted to feel about themselves with the product they were buying. The product, they were told, which might be something entirely functional like a car or a bar of soap, would make them feel powerful, or beautiful, or rich, or famous, or whatever. And, of course, it worked. For humanity it was a truly backward step, a deliberate entry into a manipulated world, where our feelings would be abused by the powerful as a means of keeping us passive. I cannot stress too much how important this is for our modern lives. This simple piece of manipulation now lies at the heart of what we have come to call 'the consumer society' and the degradation implied by calling human beings 'consumers'. The fact that we have accepted this word 'consumer' as the public description of the human being is clear proof of our current degradation and passivity and will be the source of further comment later in the book.

Another situation to consider is that of revenge. Someone is hurt. Maybe someone is killed. How do we respond? Revenge is a powerful feeling, notoriously roaming around the society like lightning seeking an object on which to deliver its charge.

Steven Pinker has researched data concerning pre-state societies, meaning those human groups that do not have a set of institutions that mediate their common life. There are cycles of revenge in such societies, as one tribal member is slain in payment for earlier misdeeds of another member of his tribe, and then another is killed in a tit-for-tat response and so it goes on. Such societies, it seems, are particularly prone to revenge killings.[12] The results of Pinker's research indicate that the formation of legal institutions in a society turns around the ability not to settle matters ourselves, but to hand our *feelings* for revenge over to the legal institution to adjudicate. There are, of course, many frustrations with any legal system and they are always only partially effective, yet they are trusted enough so that most people, most of the time, do not have to 'take the law into their own hands'. So it is that the feeling for revenge is a powerful stimulus to the development of a legal institution.

There are situations, of course, particularly in international warfare, where legal institutions fail. In the aftermath of the 9/11 attacks on the World Trade Center, there was a profound need to 'settle the matter', in terms of getting revenge on the perpetrators. Yet, as ever, revenge became a powerful and untamable force in the societies that felt violated by the attack. Leaders in the US and UK felt that 'something' had to be done, and let loose a ferocious attack on Iraq even though there was no evidence to link it to the atrocity. They tried to assure themselves of the legality of this attack by recourse to the United Nations and there was much pontificating about international law, but in the end the UN betrayed its weakness in failing to effectively prevent this disastrous political move. As I write this in 2016, Iraq and the whole surrounding area is suffering terribly from civil unrest that derives from this failure.

The Reality of the Collective

I am aware of a personal anxiety around community life that has been with me now for several decades.[13] Individualism seems to have taken hold of us to such a degree that we have little sense of solidarity in our various communities. Primary places of community remain the family and the workplace, but even these, in many instances, are poor in quality and superficial in their emotional depth.

Human communities arise very simply and sustainably from our nature as social animals. Stability is an important condition of their formation. The web of social relationships that constitute a human community needs to be nurtured by repeated interaction between its members. Walk a dog around an area for ten years and you will get to know people. Spend time at a church or a club and you will most likely find friends. That sounds obvious, but it does not seem to be obvious to policymakers today. Beginning with Adam Smith's despairing dictum that "man is of all luggage the most difficult to be transported,"[14] the modern economic system has set itself against stability and has systematically broken up communities in the name of the free market.[15] The people engaged in coal-mining in the UK may not have enjoyed the process of coal-mining, but they did appreciate the communities that had grown up around the pursuit. It was the *communities* that died in the name of modernization. An economy built on 'liberty' and 'choice' sounds great, until you realize what it is *not* valuing, like the commitment to a set of relationships, the care and loyalty that arise between people, the love of common celebrations and rituals. We have created a society where few people want to put in the commitment to stability. We would rather fly around the world on holiday, move house frequently and pursue our leisure on an individualistic basis. And it all means that community life cannot form or be maintained. Yet community life is fundamental to our sense of well-being and is part of that experience of life that is beyond words.

One important reason for the importance of community life is that the communities of which we are a part provide a 'holding context' for our emotions. They are a place where we can work out how we feel in relationship to how others feel. As such they provide for an implicit negotiation about well-being, justice, peace and all the other things we come to value.

Our present culture, for example, might be said to be emotionally childish. Consumerism works to stimulate an immediate reward system. This is the sort of emotional maturity we expect of a baby. In a consumer society delayed gratification of our desires is seen to be a 'problem', to be met by an adult version of a tantrum, rather than a normal part of an emotionally mature life. Similarly our demands for 'Freedom' can often be thinly veiled constructions of an ego that refuses to bow to the proper constraints and concerns of the collective. Social animals have to live in a collective context. There may be occasions where it is appropriate for individuals to be solitary, but the life of the community remains the abiding reference point for social animals. As such their individual lives must, to some degree, be regulated and negotiated in relation to the communities of which they are a part. The liberal agenda of Western culture does not properly recognize this. John Stuart Mill taught a version of morality whereby, we should all be free to do what we choose so long as we do not do actual harm to others. This phrase 'actual harm' is not normally taken to include the more subtle aspects of relational damage with a social system. So freedom, on this definition, becomes destructive.

There are many ways to describe a family. You can have mechanical descriptions based on the sharing of genes. You can have physical descriptions around who was born to whom, or you can see it all in terms of simply caring for others. All are true, of course, but different types of description illustrate different things. I have come to see families as something like a dance. I am fortunate to have a large family, with my father still alive, three sons, daughters-in-law, five grandchildren. So we are something of a tribe. And I have researched our ancestry, so I have a sense of the deeper connections that have informed who we are and how we have come to be. I see the family as like one of those community dances, like a barn dance. Everyone is up, the music is playing. Around the edge are the little ones, going this way and that, shadowed by their taller parents with a caring eye, moving more formally as they move with the music of the times and their opportunities. Nearer the center are the older ones, who move more quietly but full of grace and memory. Every now and then there is a special dance and all gather around one family member for a birthday dance, or an anniversary dance. Once in a while there is a death in the family and a different type of dance ensues. Reflective, tearful and full of memories, the family gathers. That member is no longer there to be seen, but very much remembered and so still strangely participates, as their memory is held in stories and places.

The family is more than any one member and more than the simple trajectory of individual birth to death. The family is a true collective. Our family existed before us and will exist after us. It is a bit like the ancient philosopher's conundrum about a ship, which was gradually replaced year after year, plank after plank, as bits needed mending, until the day came when there were none of the existing planks left. And the philosopher is worried about whether or not it was still the 'same ship'. Of course, the answer depends on your perspective. Do you see the whole thing as an entity in itself or do you see it just as the sum of its parts and no more? I think the family is a collective reality that goes on through the generations in a real sense. In other words, it can be experienced as a collective, in a way that cannot be fully captured by the definitions of words and language.

The German word 'Heimat' was the title of a series of films based on the life of a fictional village in the rural part of Germany, known as the Hunsruck. It focused on the lives of members of one family as they lived through the period from the First World War until the 1980s when the films were made. As a sustained piece of storytelling it is a magnificent achievement. But it is more than this. The story is told so as to draw out the feelings for a community that is struggling with enormous changes in culture, brought on by the two world wars and by the great forces of modernization and urbanization that were impacting the family at the center of the tale. One member of the family, struggling with the aftermath of his First World War experiences, simply walks away from his family in the Hunsruck, leaving his wife and children without explanation, and goes to the United States, where, over the years, he comes to found a substantial company. In his last years, he returns to the Hunsruck. He has no explanation for why he left, but he clearly mourns leaving. He left the 'Heimat', the community and people of his belonging. By contrast, one of his sons comes back from the Second World War and sets up an optical factory in his home area. The son loves his business and knows every person on his staff of 200, but then his competitors come and offer millions to buy him out. He is caught in an agony. He knows that if he sells, the factory in the Hunsruck will close and his staff will be either spread all over the country or laid off. He resolves his internal struggles in a poignant scene, where he announces to all the staff, gathered in a meadow, that he will not sell the business. He senses the importance of the 'Heimat', the place of home and belonging, the people to whom he owes a loyalty. This story captured the hearts of the German people when it was first shown. One estimate was that up to half of the population watched one or more episode. It was resonating with something that was being lost.

Yet there is something else to say about community. While community life is a necessary part of our existence, in certain situations it can become deeply oppressive. The rural life characterized in 'Heimat' is surely an example of a community where the social life might be so settled that dominant characters are left unchallenged and diversity of opinion is repressed. For this reason I think it is true that we all need to belong to more than one community. These might be family, workplace, club, political group, church or whatever, but the advantage of more than one community is the ability to escape the watchful eye of some of those very near us, who might otherwise dominate us and so explore different aspects of ourselves and different perspectives on the wider society. If one community dominates us, as in the religious sect, or the totalitarian state, or the gang master's world, or even the overbearing family, then we are potentially in trouble. The reason that cities are so attractive to human beings is probably that they present a set of diverse, overlapping communities and are therefore a place of creativity and opportunity that smaller communities cannot provide.

Why Is This Chapter Important?

This chapter has shown that the origins of our community life are in feelings, not in ideas. For the last few centuries our communities and institutions have been in thrall to a system of thinking based solely on evidence and reason. As a result we have lost touch with the importance of our feelings in shaping society. Our feelings have been so regularly pushed aside in the public sphere that our policymakers no longer *feel* the world as they should. It turns out that our feelings are the true origins of our sense of fairness and our pursuit of peace and all derive ultimately from our empathic imagination, our ability to imaginatively enter the feeling world of another.

If feelings are denied a public role in the shaping of societies, then their power is not part of public discourse and the way is open for a subgroup, like the advertisers and the businesses they serve, to subvert the whole society by manipulating people's feelings. Our present 'consumer society' is one such manipulated society and, as a result, we have been rendered passive.

We have also shown that feelings lie at the very roots of our sociability and the formation of our communities. Our quasi-scientific culture has found the very existence of 'community' an enigma as its remote genetic justification has treated us all as units of individual self-interest. There is little doubt that we are going to need a renewal of solidarity among peoples across the world and a sense of connectedness with all that lives, if we are to face the challenges of today. This can only come from a renewed commitment to community itself, as the essential holding context for negotiation of our feelings and the discovery of a path to a more sensitive existence.

Later chapters will show that this failure to appreciate the world of feeling leads to other things. Feelings are the root of our value system and so our morality. Feelings are a necessary complement to scientific discovery. Feelings are the one vital route to the appreciation of the natural world. Without an appreciation of our feeling world we are really quite lost as human beings.

Chapter Three

The Quest for Truth

In a Nutshell

In this chapter I am proposing that there are two quite different types of knowledge accessible to human beings. They are scientific knowledge and narrative knowledge. Both are a vital resource for our well-being as humans and both must be respected by any sensible society. The tragedy of our current position in the Western world is that we have tried to build the whole of our civilization on only one type of truth and so find ourselves, at this crucial point in history, lacking the resources that we need to react to our situation.

Scientific Knowledge

Type-1 knowledge – commonly called 'scientific' – derives from a process of observation, theory and experimentation as applied to the constituents of the world and their interaction. Knowledge grows by carefully defining the conditions of experimental observation such that the same observations can be made by any person at any place and time. Such experimental results are used to build, test and refine a scientific body of knowledge, whose certainty and definition gradually increase as more data accrue and different fields of exploration interact. The method depends on scientists attempting to be 'objective', removing their own feelings and motivations from their study, so doing experiments that anyone can replicate, and building knowledge that is agreed by all. This thinking will be familiar to most of us, but we may not appreciate that the heart of science is not theory, but experiment. Science attains to truth simply because its claims can be demonstrated. I was once a biochemist and fortunate enough to study for a doctorate in a world-class laboratory. My supervisor, Professor Alan Williams, whom I remember with some awe, would often interrupt us students while deep in some theoretical discussion with the words: "Just do the experiment." That was how he thought and that was why he was so good at science.

So what of the theory? The theories of science begin with an act of imagination. The heart of it is to find the simplest possible explanation of the new experimental result, taking into account all that has so far been demonstrated in the field. From such a process each scientific discipline creates its own body of knowledge and takes its place within a hierarchy[1] of overlapping fields of research, beginning with physics, which studies the most basic properties of matter and energy, through to chemistry, which looks at atoms and simple molecules and their interaction, to biochemistry, with enormous macromolecules, cells and complex metabolic pathways, on to biology, ecology etc. Each field of knowledge creates a theoretical world that is *consistent* with the properties of those preceding it in the hierarchy (e.g. chemical interactions obey the laws of physics). This ties science together as one discipline. Yet each field is unique in its own way, particularly because its theoretical world is *not fully predictable* from the levels that come lower in the order of complexity. So, for example, the informational properties of DNA cannot be predicted simply from chemistry. They arise as a new property of the biochemical system. This sort of hierarchical scheme was first articulated by Arthur Peacocke in 1979. A fuller treatment would include other scientific disciplines like geology and climatology, with their own overlapping disciplines preceding and following, but the principles of this type of knowledge remain across the board.

The role of theory in science is to suggest the next experiment, not to claim truth in and of itself. Karl Popper made the powerful claim that science actually makes progress by a process of 'falsification'. He said that the important thing about a theory was what it predicted and that a useful theory must always be falsifiable. That is, you should be able to define the experiment that would prove it wrong. Since that time, others have pointed out that theoretical advance can also be made the other way round, i.e. a theory is put forward that predicts an experimental result, which is then shown to be correct. The story of the Higgs boson is an obvious example of this, where Peter Higgs provided a theoretical understanding of dark matter, predicted a new particle, then a huge accelerator was built to demonstrate it.[2] The bottom line is that however you conceive of the interaction between theory and experiment, science is actually built on experimental results.

It is tempting to think, on the basis of such schemes, that all scientific knowledge is built on physics and therefore it is in physics that the 'ultimate' answers to the make-up of the universe are to be found. This sort of thinking has led to a popular fascination with particles and energy, as if they will explain everything. Some even dubbed the Higgs boson the 'God particle' for this sort of reason, even though Peter Higgs thought this name to be ridiculous. The truth is that each of the disciplines that form part of the scientific endeavor work as semi-separate enterprises. The jump from one level to another, such as chemistry to biochemistry, can be defined to some degree in terms of complexity, but, at each level, the new complexity gives rise to a new system of relationships, which can be, and needs to be, studied as a world in itself.

So, for example, cell biology lies somewhere in the hierarchical scheme between biochemistry and biology. Most of what we call 'living systems' work as a collection of cells, equipped with an array of differently functioning components. Cell biology involves the study of how each part of a cell works together to make a living system. As such it is necessary to understand the functional roles of each component, the nucleus with its DNA, the cytoplasm, mitochondria etc. Each of these components consists of complex arrangements of giant molecules working together to perform a particular function vital to the life of the cell, but *the way that they work together* could not possibly be predicted from simply studying the giant molecules of biochemistry. This knowledge is particular to cell biology. So it is that the way each discipline of science works has a particular focus and develops a body of knowledge that only makes sense in that context. The story of the Higgs boson, for example, adds nothing to the discipline of cell biology.

This means that knowledge in science is inextricably multi-faceted. There are different kinds of knowledge arising in different fields of study and one set of knowledge cannot be predicted from the other. The holy grail of science is that we might eventually come to understand each scientific discipline well enough that we have a more-or-less complete description of the material world.[3] Such a quest is intoxicating for humanity. It seems like at last we might have real power over our human condition. It is this promise that has galvanized our institutions in the West for the last two-hundred years. Scientific atheists campaign that we should make such knowledge the basis for everything about human society. But this attitude raises a major problem.

The Great Mistake

In the fifteenth and sixteenth centuries there were a set of European wars outwardly based on religious difference, but also involving of all sorts of political allegiances and conundrums about power. Philosophers and scientists living through this period railed against religion because it was used to justify terrible bloodshed.[4] Francis Bacon, one of the founders of modern scientific method, dreamt of science as a way to put aside all this religious hatred and focus on a whole new way of understanding.[5] For Bacon this was a chance to build afresh and found an experimental science on which the whole of life could be based. Bacon was not only profoundly against religion as it was expressed in his own culture; he saw science as giving human beings mastery of the Earth. The subtitle of his *Masculine Birth of Time*[6] was *The Great Restoration of the Power of Man over the Universe.* He spoke of subduing nature, "to bind her to your service and make her your slave." He spoke of scientists "uniting to storm and occupy her (nature's) castles and extend the bounds of human empire as far as God Almighty in his goodness shall permit."[7] So, for Bacon, science was a totalizing mission, to displace religion and build knowledge in every area of life around the scientific method that he described. More recently the physicist Max Planck summed up this position with the words: "Experiment is the only means of knowledge at our disposal. Everything else is poetry, imagination."[8]

Such thoughts and attitudes prevail today in the pronouncements of campaigning atheists such as Richard Dawkins and AC Grayling. The basic error made by these thinkers and their followers is to attempt to apply the method of Bacon and other philosophers of this period not only to the natural sciences but also to the humanities, or, as they were dubbed, the social 'sciences'.

The problem with this method as applied to the humanities can be revealed by noticing that scientific study not only gives us a way of perceiving reality that has proved enormously powerful, but that it also 'does something' to the people who live by it. The method of science demands that the investigator deliberately distances themselves from aspects of their own existence. They need to do work that does not depend on their feelings that day. Their experiments cannot depend on what they had for breakfast, their relationship with their partner, nor anything on the subjective side of their lives. They are deliberately raising themselves above their feelings to a world, as it is understood, of pure rationality, a certain world of 'what is' to which, they demand, we must all ultimately give credence. If this sort of attitude is applied to the study of human society it has tragic consequences.

More than One Way to Approach the World

As reported in Chapter One, the psychologist Ian McGilchrist has demonstrated convincingly that there are at least two basic ways that human beings can approach the world around them. There is the scientific method, which provides focused attention on material aspects of our existence and is incredibly powerful in the understanding it can generate and technologies it can create, but there are also holistic means of approaching the world, which are, in a sense, prior to science and which attempt to express our whole experience of life, including those aspects of the world that science deliberately avoids. This latter approach to the world, described as 'poetry' in Max Planck's definition of knowledge, is not the fantasy that he implies. It is an attempt to integrate and express the whole realm of feeling, evidence, relationship and process and is a vital component of our lives.

Another way of putting this is to say that the human search for truth cannot be said to be complete until it has dealt satisfactorily with consciousness. In the sort of hierarchy of knowledge that has been alluded to, consciousness arises at the top level of complexity and development. Out of the interrelationship of all living things has arisen the phenomenon of consciousness. Yet how should we think about consciousness? Conscious awareness is clearly present in what we call 'higher' animals and is especially well developed in human beings. In this sense consciousness could be said to be an emergent property, like other such emergent properties in the scheme of knowledge, and its expression is *consistent* with those realms of knowledge below it. So, for example, we know that our consciousness is dependent on the animal body and the brains out of which it arises, but it is clearly an emergent property, whose activities are not fully predictable from the neurological study of the brain.

And yet, it is not adequate to simply place consciousness on the top of a hierarchy of scientific disciplines, because the emergent properties of consciousness make it a *whole new type of thing* whose study takes us out of the realms of scientific method and requires a different approach to knowledge. The experimental methods of science are only applicable to aspects of our material existence that are open to study. The issues with consciousness are that it is both irreducibly complex and, at least partially, hidden from view. Recent work with brain-imaging has elicited considerable interest as, for the first time, we have been able to see different parts of the brain light up when they are activated in a particular emotional state, or whatever. The scientific impact of this is considerable, but this approach can never produce the sort of understanding that we need for meaningful action in our lives. In the general run of life, if I want to know what is going on in the social world around me, science is of little help, because there is little or no material evidence of what people around me are thinking or feeling.

Knowledge related to the outworkings of consciousness requires us to recognize two crucial facts. Firstly, for social animals like ourselves, consciousness is, at its roots, a *social* consciousness. We are aware, from the very first moments of our lives, of a network of relationships consisting of other social beings and we live in response to that network of relationships. Such networks are extraordinarily complex and motivated in part by deep drives that are present in social animals like ourselves and which have been discussed in Chapter Two.

Secondly the human being needs to try to understand what is going on around them through the use of their own *empathy* and *imagination*. We understand why people might be behaving as they do because we know that feeling ourselves. This leads into a description of an approach to knowledge that is built not on the world as object, as in science, but on *the world as perceived as subject*. In this method we deliberately invest our own feelings, beliefs and commitments into our description of the world.

This holistic approach to understanding the world finds expression, as we have already seen, in music, the arts and poetry. Yet there is also another way of negotiating truth in a holistic way. It is a method that has found expression in all that we call 'humanities', and in one sense it embraces science and is prior to it.[9] It is the method of storytelling, or 'narrative', in its more formal terminology.

The Narrative Method of Seeking Truth

Type-2 knowledge – referred to here as 'narrative' knowledge – is based on our being able to imagine the thoughts, feelings and motivations of others. We use it whenever we need to understand human behavior, or the behavior of other animals with complex minds.

To begin with an example of something slightly 'other', recall a time you may have listened to David Attenborough describing the behavior of a group of chimpanzees. You will recognize that he does not make a simple scientific description of their behavior, which would be both boring and fairly obvious, but he enriches his account with an imaginative process, attempting to enter their world, bringing to bear all that he understands about chimp culture, then telling a story about why it is that this or that animal is behaving as they do. The way that he does this is by recognizing that chimps and ourselves share some of the same set of feelings and motivations.[10]

In a similar way, in the human context the heart of the narrative method is the attempt to imaginatively enter into our inter-human relational world, using our own feelings and motivations, and so to create a narrative about why people are behaving in the way that they are. The most obvious place we do this is in the creation of our story about ourselves. As we go through life we have to try to make sense of it, not just in terms of what happens to us, but why these things occur and how they make us feel, and what these other people around us might be thinking and feeling. We also try to map this personal story onto the narrative of our wider community and culture. It turns out that we answer all the big questions about our lives in terms of our story. For example:

Who am I? – questions of personal identity. In our very fluid societies of today many are beginning to sense the comfort of knowing our roots. Programs like *Who Do You Think You Are?* interact with a burgeoning ancestry movement as people want to know their origins as part of the story of themselves.

What is happening? – as we are confronted with the world of our senses we need to make sense of our experience. Some things we can make sense of through science, but the whole social world requires a different method. We construct stories to make sense of it.

Why are you doing that? – we all use story to give reason for our actions or the actions of other people. Courts of law illustrate the importance of this question in understanding why people do what they do. If a motive can be established then a more compelling narrative of events is constructed.

With whom do you belong? – this is the question of community identity. Human beings are social animals and our group life is encouraged and enriched through the telling of shared stories. We might naturally think of our family, local community or nation in this context today. Most traditional cultures had their storytellers, who at one level were great entertainers but they also had a more serious role. They held the stories of the culture. They knew the history, myths and legends which made sense of life for that community and at public festivals there would often be a 'telling' of these foundational stories so as to enrich and affirm the community life. The 'Ceilidh' tradition in Ireland is a good example of this.[11]

What matters to you? – all stories are undergirded by beliefs, values and commitments. This is true even for those of us who are not explicitly religious. When you listen to someone telling a story about today's world, it is not long before you get a sense of what really matters to them about it. These commitments play a vital role in filtering the details of life and ordering behaviors according to a moral sense. This area will be examined in more detail in the next chapter.

It is fascinating how powerful and attractive personal stories are to us. They are a great way of communicating, even across cultural divides. Motivations, like the ones listed in Chapter Two, are the raw material of personal stories. They are their passionate heart. They are the reason that stories grip us so powerfully at an emotional level and also why they translate across cultures. So, for example, examine any edition of a tabloid newspaper and see how many of the motivations in the charts of the previous chapter you can identify. The editors of the tabloid newspaper know how attractive storytelling is. It draws us in, fascinates us and outrages us with the lives of other people. This is also true of our classic fictional tales. Shakespeare may have been written in antiquarian English and in a society very different from today, but we can still recognize the envy of Duncan in *Macbeth,* we can laugh at the love that blinds the eyes in *A Midsummer Night's Dream* and feel the tensions around the warring gangs and romantic love in *Romeo and Juliet*. It is the motivations that are our primary points of recognition and engagement. Or, if we take some other stories: we can still feel the struggle between Mr Darcy and Elizabeth as he declares his love for her in *Pride and Prejudice* yet battles inwardly with his own status anxiety about her being not quite from the right position in society. We feel a love for the suffering boy *Oliver* as we feel for the young everywhere and for the weak and the marginalized. We enjoy the tussle between Hugh Grant and Julia Roberts as love overcomes the barriers of status and celebrity in *Notting Hill*. We are inspired by the *Braveheart* who avenges the death of his beautiful secret wife and whose personal struggle is translated into a very political and bloody power struggle with the English, while he contends with the competing obligations of other Scottish tribal leaders. We are fascinated by *Borgen* as a political tale that involves full and rounded characters delicately balancing their private lives with their political responsibilities. Every story can be read from a motivational viewpoint and to do so reveals its power for the reader, listener and viewer.

Narratives and Public Truth

A similar narrative method should be recognized to be at the heart of all those disciplines we call the 'arts' or 'humanities'. These include:

History
Politics
Law
Sociology
Psychology
Economics
Religious Studies

The common denominator in all these fields of study is an attempt to make sense of human society by constructing a narrative that draws together observed behavior with our imagination of the inner lives of people. Such narratives are holistic theories of life. These theories are not more trivial nor any less important than science, though they are much more uncertain. Narrative knowledge is a necessary part of any human culture and forms the basis for all its most important decisions.

Let me explain how this works with a few examples.

An example from history:

A historian attempts to get at not only what happened, but why it happened, telling a story about principal actors in their narrative, their lives and the pressures on them. They will research the sources available, consider what sort of people produced the sources and what they know of the culture in that period, and so come to a view of how to express a 'true' reflection of what happened. This process requires the historian to subjectively 'enter the world' of the period under study, pouring their own imagination and empathy into the study. Christopher Hill, for example, wrote an inspiring history entitled *The World Turned Upside Down*, which focused on radical movements in the seventeenth century. It questioned property laws, the enclosures and the oppression of the poor. He tells, for example, of the Digger movement and of Gerard Winstanley and his attempt to build a radical community based on common ownership. The book is widely recognized as an historical masterpiece, and it is made all the more effective by Christopher Hill's own commitments, as a radical scholar and former Marxist. The role of such pre-formed commitments in storytelling will be considered more fully in the next chapter, but the idea of a 'neutral' history is both impossible and doomed to failure, because the very act of storytelling demands that the teller personally 'enters into' the situation and responds to it. The storyteller can only do this using their own emotional repertoire and the commitments that give it shape. It turns out that it is better for a historian to know the commitments they are bringing to the evidence than to pretend neutrality while smuggling in their own subconscious and unexamined approaches to the text.

An example from politics:
Political method can be described in terms of a set of narratives being told to the public in order to justify actions. The politician may make it seem like they are acting out simple policy, or responding to the dictates of the economy, but underneath there are deep commitments being brought to bear and the whole process can be understood as the construction of a narrative in which a politician is saying to the public, "This is the way I see this thing, enter my world and approve my action." One terrible example of this concerns the war in Iraq, where then-Prime Minister Tony Blair told us a story about weapons of mass destruction, about the evil nature of Saddam Hussein's regime and the rapidity and ease with which any struggle would be settled, and then called on us to approve his decision to go to war. Many of the public did not believe the story, but we were assured that there were secrets that we had no access to and were hard-pressed to refute it. Two-million people demonstrated on the streets, but the seemingly inevitable march to war continued. The repercussions of this are still being felt. There were elements of this story that were open to inspection, and evidence, now known to be inadequate, was brought forward to justify it. Yet underneath the story, and informing it, were clear and powerful commitments that related to the ego of politicians and our relationship with the United States.[12] So here was an example where the political process seemed to fail the people. There was no way to challenge the dominant story.

Truth and Right are Blended Together in Narrative

It is apparent, from this description, that the narrative process involves a subtle blending of truth and right judgement. In our assessment of all narratives, we need to do two things. We will weigh the facts as presented by science or other clear evidence; and we will also pay attention to the motivations of the key actors in the drama together with their beliefs, values and commitments. This means that the judgement we need to come to is necessarily about both what 'is' and what 'ought' to be. It means that truth and right judgement must hang together in a coherent narrative that provides us with our judgement about the matter in hand.

So, for example, I might conclude what was true about the invasion of Iraq was:

1. that there were no weapons of mass destruction.
2. that our Western leaders were corrupted in their perception of the situation by their own egos.

This would lead to a judgement about the prosecution of the war that may be accorded 'true' by others around, but might also simultaneously be judged 'right' by these same people on moral grounds. Narrative-type knowledge inevitably has to embrace both truth and right and must involve us taking a view about the beliefs, values and commitments of others.

It is clear from all this that any account of politics, or history, or economics, or any other study of human social behavior, is always partial and uncertain. It functions in society as an *appeal* to truth. This means that there needs to be a process in these disciplines of considering alternative narratives as part of the wider search for truth. There is always more than one story that can be told. Such alternative narratives need not be mutually exclusive but can offer complementary or contrasting perspectives that enrich the whole account. This is, to my mind, the most important justification of a democratic process in human societies. Leaders tell stories and the people judge whether or not they are true and right. Democracy gives value to everyone's interpretation of the world and the contrasting narratives they may hold.

Why does this matter? Some academics would say this whole process is so difficult that it is best ignored. We should simply accept that you cannot get at truth in the humanities. That is easy for academics to say, but it is impossible for them, or anyone else, to live by. We need to act in the world and we all act out of those narratives that we take to be true. So, the politician *has to act.* That means they have to commit themselves to a story about the matter at hand. And the public need to respond to that story. It is the same in our own lives. We construct a story about our lives, not as some pastime, but because it really matters. Psychologists tell us that without a story we are lost. Whether we do this story-telling consciously by means of a diary, or in conversation with our family, or whether it is a largely subliminal and unconscious process, nevertheless we are constructing a story and living out of it. So it is with all the humanities. They are the basis for action in a culture.

A Summary of the Two Approaches to Knowledge

I have outlined two quite different approaches to knowledge, each of which is appropriate in its own sphere. I am proposing that only if we embrace a clear distinction between these two types of knowledge can we adequately assess their respective claims to truth.

This illustrates how different they are:

Type-1 knowledge – scientific:
based on study of world as object
deliberate dispassionate engagement with material world
using reductive analysis
working for the simplest possible explanation

based on repeatable experiment
subject to peer review
commanding truth

Type-2 knowledge – narrative:
based on study of the world as subject
necessary engagement of imagination and empathy
working for holistic narrative
recognizing inevitable complexity
requiring alternative narratives
using beliefs, values and commitments
appealing for recognition of truth

Scientific truth claims arise within the scientific community and are tested typically by peer review. Those schooled in the discipline assess the evidence presented by their colleagues, comment on it, and argue for further definition, experimentation etc. as appropriate, in order that the study makes clear conclusions with the highest possible degree of certainty. The scientific community and the wider community together then have an *interpretive task* in relating this knowledge to the rest of life. *This must involve creating a narrative that situates this experimental finding within the context of history* and explores its relevance to our future. You could say this involves moving from an 'is' (the body of scientific knowledge) to an 'ought' (this is what must be done in response). The 'is' stage is a scientific quest and must be assessed on that basis. The 'ought' is a matter of public policy and involves the creation of a narrative that must engage with moral priorities as perceived by the culture.

Why Is This Chapter Important?

Over the last few hundred years we have been through two distinct phases. The first 'modernist' phase was marked by a huge confidence in science as the answer to all our problems. Social theorists, like Jeremy Bentham and John Stuart Mill, sought to design society by drawing on scientific method and claimed a new certainty for their results. In a similar way, philosophers like Immanuel Kant and John Rawls tried to derive moral guidance from a standpoint of pure rationality, while historians attempted to convince themselves that they too were working simply from evidence and reason.

The second stage, characterized by 'postmodernism', was to realize that all these thinkers in their different fields had failed. People like Jean-Francois Lyotard went for the jugular in the whole body of modernist 'knowledge' and showed that even science was not as value-free as had been claimed.[13] Scientists are part of institutions and they need funding. Evidence was produced to show that the areas of investigation that scientists choose to work on are strongly influenced by whether or not they can source funds and that this funding process can compromise their conclusions. So, for example, the research focus of medical institutions can be slanted towards the interests of the drug companies[14] that fund the research, agricultural research influenced by international seed or fertilizer/pesticide providers, climate-change science influenced by big oil etc. The data in this area is shocking and some of it will be considered further in Chapter Five.

Likewise, in the humanities, the attempts to justify morality by simple rationality were also a complete failure.[15] Utilitarianism, the theory that one can justly order society around the greatest happiness of the greatest number, was clearly flawed. Historians were similarly shown to be biased by their own commitments. Overall the failure to justify truth on purely logical and rational grounds then gave rise to a fashion of academic skepticism whereby universities expended great energy in *preventing* anyone from approaching truth in the social sphere, appreciation of literature was emptied into meaninglessness, and all attempts to build knowledge were quashed.

This very brief sketch outlines the confusion that lies at the heart of Western society today. We simply must recover confidence in both types of knowledge if we are to survive the crises that lie ahead for human society. We need to regain a proper estimation of the authority of science, in regard to such as climate change. We also need to recognize the essential part that narrative engagement has to play in the establishment of good policy and in the social arena generally. Only then will we be able to make good decisions together. Chapter Four will show that narrative failure lies at the heart of our current environmental crisis.

Chapter Four

Necessary Commitments

In a Nutshell

We like to think we are pretty rational people, going about life, making all sorts of decisions based on weighing alternatives, choosing the best path and carving out our future by recourse to reason. And yet the truth could not be more different. This chapter recounts evidence showing that most decisions that we make in life are not the free choice we might pretend they are, but are actually formed from deeper commitments that give shape to our lives. These commitments form a hidden world within us. Frequently unexamined, they are the source of immense energy and passion, a passion that can drive us to make extraordinary sacrifices, to work without reward and to unite with others in the formation of communities. One of the most crucial dynamics of any society is the use of story to mediate between members with different deep-level commitments.

Discovering a Hidden World within Us

Every social animal has to make rapid decisions about whether to 'approach or avoid' another of their species coming toward them. They need to be able to spot who is in their group and who is not, what their demeanor is, whether they will fight etc. And they need to do it fast. The accuracy of this ability will determine the animal's survival. For an animal all such decisions are emotionally laden. The mouse in the field may glimpse the shape of a bird of prey in the sky, feel a deep fear, and then bolt for cover. The emotional response helps and works with the physical one, sharpening the senses and galvanizing the body for flight.

Since the Enlightenment, many of our philosophers have dreamt of a world founded purely on reason and defined the truly great human beings as those who can rise above their passions into a world of abstract reason.[1] It turns out that this is a dangerous illusion. It is dangerous because our moral sensitivity arises from the world of our feelings, not our reason. It is an illusion because none of us actually works that way.

According to the psychologist Jonathan Haidt, we actually make commitments about life based on a sort of 'moral intuition' and these commitments help us form our first response to the situations we face. Haidt speaks of our minds as working like an 'elephant' and a 'rider'.[2] There is the dominant part of our mind, the elephant, which carries our fundamental commitments about how to approach life, and the rider, which tries to interpret these commitments rationally to those around us. Haidt is an experimental psychologist and has proved that human minds react immediately to events around them based on an intuition that is informed by our deepest moral sensitivity. So, we meet someone, and we respond inwardly and intuitively to them, even before we think about it. Only then do we start to *think* about our response, and that thought process is usually used to justify our intuitions.

Haidt believes that the moral commitments we all work with have emotional roots and traces six lines of development based on:

Care/Harm
Fairness/Cheating
Loyalty/Betrayal
Authority/Subversion
Sanctity/Degradation
Liberty/Oppression

Each of these commitments is triggered and enhanced variously by the experience of our lives.

For example, a mother whose child endures a particularly long or difficult type of suffering will be motivated strongly by compassion and may develop a deep concern about those who suffer harm in the world, especially children.

Or a worker in a hierarchical workplace may come to respect 'the boss', relate in a subservient way and recognize the value of being in an 'in' group. Or, alternatively, if the experience is felt to be dominant and overbearing, the worker may become subversive, or embrace commitments that favor 'liberty' over 'oppression'.

So it is that our experiences of life create passionate commitments within us that fuel our lives and shape our general approach to the world. The lists of our emotions in Chapter Two illustrates how complex this process can be and how many different feelings are in play. Yet for each person our feelings, coupled to our particular experiences, do gel into a set of commitments that take us through life.

Adam Smith may be best known for his book on the wealth of nations, but he was also a very interesting moral philosopher. His book *The Theory of Moral Sentiments* shines a light on the way that human intercourse revolves around deep emotional sympathies, whereby we sense what someone else is going through and adjust our response to them so as to create a sort of emotional solidarity.

A similar approach has, more recently, been taken up by the Common Cause Foundation, who recognize two basic types of values, termed 'extrinsic' and 'intrinsic', which are antagonistic within our minds. Extrinsic values are things like:

> *wealth, social recognition, social status and prestige, control or dominance over people, authority, conformity, preserving public image, popularity, influence and ambition,*[3]

while intrinsic values include:

> *'broadmindedness', 'a world of beauty', 'a world at peace', 'equality', 'protecting the environment', 'social justice', 'helpfulness', 'forgiveness', 'honesty' 'responsibility', 'self-acceptance', 'affiliation to friends and family', and 'community feeling'. They also include values associated with 'self-acceptance' (defined as a feeling of competence and autonomy, including feeling good about one's abilities, feeling free, and having insight into the reasons why one does things).*[4]

Evidence suggests that appealing to intrinsic values creates a certain resonance in our minds with other intrinsic values, while appealing to extrinsic values similarly reinforces other extrinsic values. Socially progressive movements are often caught in a dilemma with their communications and can be tempted to think that appeal to extrinsic values will attract more donations and recruit volunteers most powerfully. So, for example, they might offer a 'donate and win' lottery to encourage donations or offer some special status to high net-worth individuals. Common Cause gives the lie to these practices, proving that worthy causes must primarily appeal to intrinsic motivations if they are to have lasting impact. This thinking has already proved a useful resource for the charitable and environmental sector.

Much of this process of values assessment is hidden from our conscious awareness. We just do it and some of the development is way beyond our control. There is evidence, for example, that juvenile mice that have been nurtured by a good mother, as measured by the mother's diligent attentiveness, become good mothers themselves, while the reverse is also true with bad mothers bringing up bad mothers.[5] In the human context parental weaknesses might similarly be passed on and can result in not only poor child-rearing, but a broader diminution in social sensitivity among offspring. There is little doubt that the strengths and weaknesses of a child-rearing regime will change the spectrum of commitments that a child might grow up with.

In a practical sense the development of moral commitments gives us rules of thumb by means of which we can navigate through the massive number of decisions that we face each day as members of a society.

This is made absolutely clear in the world of politics and allegiances. Most people know how they are going to vote long before an election campaign begins and there is little that anyone can do to change that commitment. The politicians can expend enormous energy and money on a campaign, but nothing moves them. I wonder why people are so settled in their opinions? It is surely a sign of their hidden commitments about life. And the political pundits know this all too well. So the campaigns seek to identify the floating voters, whom they stand a chance of influencing.

Haidt has made a particular study of political commitments and relates the six moral commitments listed above to a political analysis. Coming from a United States perspective, he finds that their 'liberals'[6] respond primarily according to a moral sense of harm, fairness and oppression, while their 'conservatives' utilize these aspects of a moral compass, but also include a deep respect for loyalty, authority and the sacred. The data on these questions are quite convincing and prove that political allegiances are actually set by feelings and values rather than by reason.

Politics is a good example of the world of commitments, but it is not by any means the only example. If we could track our lives and the choices that we make, even down to what we search for on the internet, who we 'like' on Facebook, the shops we frequent, the sort of people we know, the communities we associate with and the schools we send our children to, all these aspects of our lives actually reflect our deepest commitments. Each one does not simply derive from a free choice made at the time, but is influenced by the set of feelings, values, beliefs and commitments that reflect our predetermined approach to the world.

Less Examined Commitments

One of the perennial problems with the world of our commitments is that we are not always aware of them. We have not examined our lives well, together with the decisions that we make and how we feel about them. The choice of schools for our children is an obvious example. Where parents are rich enough, or influential enough, to have a choice of school, their decision about schooling often becomes a moment where deep, unexamined commitments are expressed, some of which they may have inherited from their own parents. They might scan the school league tables in order to find a high-performing school, but at the same time they will also be considering the 'ethos' of the school, which is a term often used to describe the felt values that the school is promulgating. This 'ethos' can be felt in the attitudes of the staff and pupils, the sense of order and calm among the pupils, or the creativity of their work. Parents might also look for a type of school like their own, or one that is an apparent gateway to the life they would want for their child. Such choices can have dramatic effects on their children.

For example, I was sent to a minor private school. My parents had been to a private school themselves and, for them, this was something that you did for your children, if you possibly could. They made considerable sacrifices to send my brother and me to this type of school, firmly believing they were doing the best for our future. So it was as a child aged eight I found myself in a dormitory with other boys, missing my parents and without any real comfort. I learnt to survive in a society of boys my age, with a set of masters, who were both kind and threatening. My school was probably one of the more benign forms of private school and many of the teachers very well-intentioned, but I am aware now that the experience was one that led me to close down on my emotional life. At the same time the school life acted to cut me off from my local community and fostered a general sense that those 'out there' were lesser beings. Private-school education might be very good at an academic level and certainly created academic opportunities for me that might not have arisen, but it also created deep attitudes that I have spent much of the rest of my life trying to undo. It is this rarified social context that insulates the growing child from a broad spectrum of people, particularly those more disadvantaged than themselves. It is the implicit set of assumptions that subtly justifies wealth and despises poverty. It is the school magazines that celebrate the great achievers and fail to tell the more difficult stories. It is not surprising if such regimes create certain political commitments.

The private-school example leads to another idea, which is that deep values are more often caught by a process of mimicry than explicitly taught. People sense the attitudes of those around them and group pressure builds on them to conform to the dominant mores of the group. Psychologists have done some dreadful experiments with group pressure, proving that people will hurt others in order to remain in an 'in' group. History, in regard to our treatment of minorities, proves the same. Every mother also feels concern about the power of group pressure as it might affect their children. As a result they will personally worry and may express concern about any 'bad company' that their adolescent is getting into. Values are catching. People are changed by group pressures. It matters how we manage our social arrangements.

So it is that the commitments that we need to navigate our way through life arise from the world of our feelings and are translated by means of our life experience into a set of values from which we live. Good societies continually negotiate these values. For example, considering the list of feelings and motivations in Chapter Two, most societies would tend to honor value sets associated with the love of family, community and the other, and work to uphold trust and loyalty, while frowning on greed, lust, jealousy and suchlike. The way that we negotiate these things involves storytelling.

The Power of Story to Change and Develop Our Commitments

As well as the particular power of those around us as young people, we are also extraordinarily influenced by the stories that we come to feel part of. We absorb a whole range of stories from the culture around us. These may be fictional stories as well as historical and political tales and other forms of non-fiction. The important thing about stories is that they are holistic in their very methodology, wrapping up our whole world of feelings, values and commitments into a coherent narrative about social events. As such they present their appeal to us, saying "enter our world and embrace the commitments with which we are laden." Much of the attraction of the stories we embrace will turn around their values. Of course, the impact of a story on us will depend also on how well the story is told, how authentic it feels in terms of human relationships and, if it purports to be history, the way that it deals with the relevant evidence; but values will be the main attractor.

There are pivotal moments in the history of Western society. They focus on a shift in the deep-level commitments of the society. One such moment was the abolition of the slave trade. A few people in England had come to feel the problem. Many on the plantations were suffering and some were stirring towards revolt. Many years of campaigning can be summarized in a simple message, which became a slogan for the movement. It was a picture of a black slave with the phrase, "Am I your brother? Am I your sister?" The campaigners brought the stories of the slaves into the public arena. They told of the conditions onboard the slave ships. They proved the terrible toll of lives lost in transit, of rape and violence done to the captives. And they shamed the populace of the UK into a deep-level change of perspective. At that time the society had been truly split about the trade. It was 'out there'. Little was known about what was going on and, while little was known, it was easy for the smooth justifications of the 'trade' to continue. In Bristol, where I live, our churches were so embedded in the trade that they actually rang their bells in celebration when the first bill to abolish the slave trade failed to get a majority in parliament. Yet our perspective has changed so much that now there would be no voices to support such treatment of human beings. How did this happen? I would suggest that the campaign can be summarized in terms of the hearing and telling of stories. Whether it was hearing of riots in the plantations, or listening to presentations about slave ships, the people of the UK were faced with the reality of those subjected to the trade. They became real, real people who were acknowledged to have feelings like ourselves through encountering them and the telling of their stories. A very similar analysis could be made of the Civil Rights movement in the US and the ending of Apartheid in South Africa. Each of these involved an encounter with one who was perceived to be different in such a way that their common humanity was appreciated, social sympathy was engaged through powerful storytelling and the perception of society changed.

I believe that the great narrative challenge of this age concerns our relationship with the Earth. The reason we are making such an inadequate response to climate change, the extinction of species and the whole host of issues concerned with our relationship with the Earth is not a scientific problem at heart, it is about forming a narrative. We need to come to *feel* for our destruction of ecosystems. We need to *grieve* for the loss of species and we need to *recognize* ourselves as powerful participants in the great dance of life around us. Such feelings need to be expressed in narratives that unite us, move us, change our minds, evoke sacrificial lives and raise up movements of social transformation. In two centuries people may look back on this generation as we now look back on the slave-trading people of our past and wonder how we could have failed to see something so obvious. Chapter Five will pursue this issue in more detail.

Storytelling is such a powerful thing. It is truly how we negotiate the deep things of life. The Enlightenment has tried to portray the good society as dominated by reason, but no, actually the good society is pervaded by rich storytelling.

I invite the reader to consider your own preconceptions about the world and an instance when your perspective has changed profoundly. Did you meet someone who was different from you and impressed you? Perhaps someone who told you a story, or offered a way of seeing things that you have never before considered? One way we can envisage life is as a developing story. Every time we meet someone, every time we engage in conversation, we are listening to another person's way of seeing the world and touching their story. People pass on their perceptions of others in all sorts of ways, the language they use, their facial expressions and body language as they speak, and we sense this stuff at a deep level. Remember a time when you have come away from a meeting with someone and something inside has gone "Whoa, watch out there!" You have sensed a clash at the level of deep commitments. You have touched their story and did not like its tone. You may have responded by telling a story yourself that shows a different perspective. Or you may have just retreated. Such is the negotiation that goes on in any society.

There are also more formal storytelling activities that shape a society. Fifty years ago many of the children in England attended Sunday school. There they were exposed to stories from the Bible and, particularly, from the life of Jesus. Sunday-school teachers would scour the scriptures for stories that children could appreciate and would often amend or embellish the tale so as to bring out a useful moral lesson. Some of these lessons were more useful than others and many had a more moral than religious intent, but there is no doubt that they came to shape the deep commitments of the young lives that heard them. Many of the older people of today attended Sunday school in their youth and learnt a moral code that has stayed with them all their lives.

As one example of this, consider the story of the Good Samaritan. It is a simple but rich tale that draws out our feelings of love for the marginalized or unjustly treated. A man wandering down a dangerous road is attacked and left by the roadside. Religious people pass by, excusing themselves as they go, while a stranger, a member of a little-known people who were often denigrated by the original hearers of the story, stops and takes care of the wounded man. There are many lessons that can be drawn. For example: religion without compassion is inadequate; compassion must be enacted; those who you suspect, or blame, or treat as outsiders may prove to be more 'human' than your own people. These are some of the things that this story has embedded in the consciousness of the generations that have heard it. There is no doubt that it is powerful and few would deny its goodness. So it is that a society finds itself replete with such stories, stories that mark its deepest commitments and shape the lives of its people.

Alasdair McIntyre says of stories:

> *Man is in his actions and practice, as well as his fictions, essentially a storytelling animal. He is not essentially, but becomes through his history, a teller of stories that aspire to truth. But the key question for men is not about their own authorship; I can only answer the question 'What am I to do?' if I can answer the prior question, 'Of what stories do I find myself a part?' Deprive children of stories and you leave them unscripted, anxious stutterers in their actions as in their words.*[7]

The richness and diversity of the stories that constitute a culture are a good measure of its health.

Castles of the Mind

The discussion so far takes a very positive position regarding the pre-commitments with which we approach our lives and how we negotiate these through story. Yet, we will all know people whose opinions about life are not negotiable at all. We call such people bigoted, prejudiced or even 'fundamentalist'. It is interesting to consider what is going on in such people and how they are distinguishable from what has already been described. My sense is that we may describe people as prejudiced, or bigoted, not because they hold an opinion strongly, but because they are unable to consider any alternative. The stories, or even the science, that offer a countervailing argument are like water off a duck's back. They are batted back without even a moment's consideration. It is like the person has constructed a castle in their mind that is designed to keep them safe from contrary views. There is no real quest for truth. There is no real openness to sympathetic understanding. I think we have all met such people. It may be expressed in political opinion with calls to "hang Gordon Brown or Ed Balls," or a decrying of the "left-wing charade" that was the opening of the Olympic Games. It may be skepticism of climate change, distrusting the opinions of scientists, however many there are and however worthy their credentials. In a person of faith this may be coupled with disbelief in evolution, and an expectation of the imminent return of God, who will sort things out in the world with an "iron hand." There may be an expectation that the financial system may collapse or a belief that Africa can only be "sorted out" by a new process of colonialization. Such is the sort of language used by a person who has constructed a castle in their mind. It may surprise the reader to learn that I actually heard every one of these things said by the same person in a conversation lasting less than fifteen minutes!

This sort of analysis can also be applied to other types of fundamentalist, whether it is the Islamic extremist or the ultra-atheist. The same psychological traits are evident. They live in a castle of their mind. Some like to see the 'outer world' collapsing because this confirms their beliefs. They remain safe in their castle. The dangers are also evident. Those outside the castle are not accorded a fully human sensitivity. Violence can be quite close to the fundamentalist mind. It can consider the suffering of others without flinching and can even approve it, or cause it. We are rightly worried by people with such a mindset.

One way to describe the extreme and violent fundamentalists by whom Western society is currently threatened is that they have problems in telling their own story. It may be that second- or third-generation immigrants have a tendency towards alienation, feeling that they belong neither to their originating culture nor to the society in which they now live. Their recourse is to create a fictional and extreme narrative that gives certainty where all was confusion and that produces violent fundamentalism.

The Constraint of Rules and the Development of Trust

Trust is probably the single most important ingredient of a creative social order tending towards well-being. Those areas of the world where the social order is fragile tend to be characterized by gang warfare and terrible vulnerability. Yet, the secrets of developing trust are a little-understood art.

A working trust arises from the interplay of feelings in a society. Such feelings develop into accepted values, which people share and know they share. Shared values are a vital source of trust between people. Communities may actually form by a process of mimicry based on shared values. So it is that feelings and values are inextricably linked in any society.

There is a terrible tendency for a society to extend its rule base and to diminish the realm of creative trust. This, according to Ian McGilchrist, is the besetting sin of Western society.[8] Our brains have become dominated by 'left brain-dominated operations' such that we favor static representational views of society, like those provided by rules and bureaucracy, over right-brain processes that are dynamic and creative.

The struggle between rules and trust has been most obviously played out in the UK in recent years in the area of health and safety. Children's lives, especially at schools and clubs, are now so dominated by health-and-safety considerations and the fear of litigation that many activities that are useful in children's development are ruled out. Children's play is hedged around by every safety barrier and cushion. Activities are scrupulously analyzed and logged to ensure accountability. And the children feel caged by the set of rules around them.

In a similar way, few people today recognize the importance of ritual acts in community formation and the development of trust. Yet some do. I was talking recently to one of the founders of the Holzmarkt in Berlin. Starting from a trailer camp more than ten years ago, this community project on the line of the former East/West border has now mushroomed into a flourishing community venture with very serious investment, building projects and the like. Part of their success might be linked to the conscious development of rituals like annual festivals and dances around which the community has formed. The organizer remarked to me, "We make a ritual out of everything!"

The commitments shared by any community are systematically enhanced by ritual acts and celebration. This has been the great method of religions with their power to give cohesion to a society. Emile Durkheim showed that religion had an extraordinary ability to draw people together, celebrate a common story and give energy to a community's common life. Yet Durkheim suggests that this process is not actually a function of the numinous aspect of religion. It is not dependent on the particular belief. It is rather a function of commonality. It is the shared singing, dancing and repeated ritual in which the individuals lose themselves in the moment of community bonding and the celebration of common purpose.

I shall develop this theme later in the book with a proposal for a global ritual act associated with our common identity as creatures with a responsibility to live in harmony with the Earth.

Living in the Cult of Western Society

It is easy to knock the fundamentalists with their 'castles of the mind'. It is less easy to see our own captivity. There is something hidden about the commitments that rule our lives, hidden by ourselves, from ourselves. What I mean by this is that many of us live without examining our deepest commitments. Whilst we may examine the commitments that our parents pass on to us (and teenage years are a vital time for this), we may still fail to examine those enormous commitments that sneak into our lives as part of Western society.

The dark side of this came through to me in the late 2000s when I remember hearing how the police had infiltrated a band of environmental campaigners, to spy on their activities. This subterfuge seemed shocking at the time. These people were nothing other than political campaigners with essentially benign methods. So why was it appropriate to spy on them? In the unchallenged language of the police at the time, they were 'extremists', who threatened the social order. Whilst for me, they were enlightened people who were trying to make a statement in, what I thought was, a democracy. It was then that I realized how narrow the line is between upholding social order and branding all radicals as extremists to be spied on and potentially suppressed.

In May 2015 I had a nightmare. I dreamt of people sitting around the walls of a room, some in uniform, different uniforms, denoting different status. Some had the frightened eyes of the condemned. Everyone was quiet but the atmosphere was tense.

As I looked, I saw that the room was part of an enormous complex. My gaze went from room to room. In each the same silent people with scary eyes. I remember in one room, a ritual was taking place. It felt like an 'ending' ritual. A life was being snuffed out with a quasi-religious ceremony. And no one resisted. There were those who were clearly in charge, but there were no guns and no chains evident. The people were just passive and unmoving. They knew they were into something bad, but could not see a way out. Then I just said to some of the more able-looking, "Let's walk out. Now. Come with me." And once I had said it once, I had confidence to say it again, more loudly. "Come on, we are going." And the people began to rise from their beds. The captors looked on, suddenly impotent, as a new awareness dawned on the captive people. A great tide of people emerged into the sunlight.

It was, of course, just a dream. But one I thought I should pass on. I am firmly of the opinion that we need to rise up and find a new way of life as citizens in Western society. In their inspiring book *The Abundant Community*, authors John McKnight and Peter Block argue that consumers are "people that have handed over to others the power to provide for a full and satisfied life." And they state boldly, "We are not consumers… We are citizens with gifts and power to make powerful communities."[9] I like this language. Once we have comprehended again how human societies actually work, we will see our present captivity and have the opportunity to live differently. The rest of this book is dedicated to exploring this process.

Why Is This Chapter Important?

There are two ways that we can live. We can live passionately or passively. The commitments that we bring to our lives are the signs of enormous energy within us. The human being is rather like the contours of the Earth, which is shaped and formed by powerful processes seething below the surface. Every now and again a volcano erupts and great damage is done, but most of the time life goes on unaware of the great energy that drives the process.

Human beings negotiate their deepest commitments through story. Every society needs to examine the quality of its storytelling. A good story challenges us to our heart. It opens up new ways of seeing the world. It shapes our politics, our faith and our life. It can topple the most entrenched bigot. It can release fountains of joy. There is no one right story. There are always many ways a story can be told, but each, like the stroke of an artist's brush, provides a new color to the world.

I have written this chapter, not so much to show how irrational we are, but rather to point to the energy that, if we were to tap into it, could give such creative power to our lives. It is time to awake from sleep. In a later chapter, I shall develop the idea of the ground of energy within us that is expressed by our whole set of beliefs, values and commitments and try to explain the importance of personal examination of this great area of our lives.

Part II

Cross–Currents

Great wings
moved graciously
through buffeting winds
the eagle's eyes focused keenly
on the scene below

Chapter Five

The Subversion of Science

In a Nutshell

This section of the book uses the principles established in the first four chapters to illustrate the pervasive weaknesses that now underlie Western society. It looks at science, economics, politics and religion and shows how each, in turn, has become impotent in the face of the challenges that now face humanity.

This chapter shows how the knowledge accruing from science on matters of vital importance, such as climate change, has been systematically undermined. This has been achieved by a handful of rogue scientists and non-scientists who have hidden commitments relating to the free market and minimal government. They have worked with a media that is corrupted by its commitment to audience ratings and a false notion of balance and also with politicians, who find it difficult to lead on anything that relates to future good rather than present enjoyment and who fear alienating the public.

It concludes that we need to recognize the distinction between scientific and narrative types of truth and examine the inner commitments that we each bring to our descriptions of society.

Undermining Scientific Truth

The depth of this scandal became clear to me in September 2013 when the Intergovernmental Panel on Climate Change produced the latest in its series of reports concerning the science of climate change. It was a summary of the work of over 800 scientists, who are specialists in the disciplines associated with climatology.

A panel including government representatives from 32 states

had argued over 1855 comments on the report to ensure fair and accurate representation before it was published. It is a complex area of science and the report had to make predictions about future trends in climate. The scientists attempted to judge dispassionately the level of uncertainty in their predictions so that policymakers could respond accordingly.

So far, so good. The scientific element had been presented. They had established what '*is*' happening to the Earth. Now the society had to do some work with a narrative type of truth-telling in order to figure out what '*ought*' to be done in response. Terrible flaws were immediately revealed in the narrative process.

Here in the UK, a clearly orchestrated media campaign kicked in alongside the publication of the report. A set of a non-scientific, but highly influential, people related to Lord Nigel Lawson and the Global Warming Policy Foundation (GWPF) sent a wave of comments through contacts at the BBC, ITV, the *Telegraph*, the *Sunday Telegraph* and the *Sunday Times*. These involved Nigel Lawson's son, Dominic Lawson, and other associates of GWPF, including Lord Ridley, who is brother-in-law to Owen Paterson, then UK environment secretary, who also, in his turn, passed on misleading information at a fringe event of the Conservative Party conference.

This attack on the report included an attempt to personally discredit the scientists of the IPCC as a 'politically motivated pressure group' and their report as a partial, misleading and unrepresentative representation of the scientific data.[1]

I was so dismayed by this turn of events that I looked further into the problem of genuine scientific conclusions being undermined. Naomi Oreskes and Erik Conway, in their extraordinarily detailed book *Merchants of Doubt*, have researched this type of phenomenon as it has been practiced over several decades in the United States.[2] They have discovered that a very small number of influential people there have worked on a succession of issues like passive smoking, acid rain, the ozone layer and climate change. In each case their intention has been to neutralize the impact of the science so that public policy is not affected. Several of the people involved in this attempt at scientific obfuscation share a similar story. Frederic Seitz, Bill Nierenberg, Fred Singer and Robert Jastrow were all part of a pro-*Star Wars* lobby in the Reagan era. This was a campaign for a great shield over America's airspace, which arose from deep fears about the Soviet Union and its intentions. Such campaigns can become commitments, deep, unexamined commitments, that spill over into other things. As the Soviet threat subsided, it seems that this group of people saw the next threat coming from environmentalism and cast environmentalists as the enemies of freedom. They saw environmentalism through the same lens as their anti-communism and feared a coming 'socialist' order for the world. Part and parcel of this belief system was an unshakeable belief in the free market, akin to what George Soros described as 'free-market fundamentalism' and which I have termed a 'castle of the mind' in the last chapter. This desire to maintain the freedom of the market is the deep logic behind all the campaigns against passive smoking, acid rain, responding to the ozone layer etc. Nothing should be allowed to interfere with the free flow of the market. Anything else was an offence against the true liberty on which progress was dependent. It appears that it is precisely this pre-formed commitment that is also calling the tune on the campaign to confuse the science about climate change.

We might argue, of course, that people are entitled to their views and that proper debate in the public arena should allow contending opinions to be heard and weighed. That, of course, is something I most strongly believe in. What is less admirable, however, is when people deliberately obfuscate the science that lies behind these debates. Oreskes and Conway show that time after time, the integrity of scientists has been challenged by this cabal in the US and the science itself has been deliberately misrepresented so as to deliver false conclusions. They cite evidence that 52 out of 56 books published on an 'environmentally skeptical' theme in the 1990s originated from one of the right-wing foundations through which these influencers work. Such books are not subject to peer review by scientists in the field like conventional scientific publications. So they can really say what they like and claim anything they think they can get away with in respect of their readership, which is typically a non-specialist, general audience. Much of the funding of these foundations derives from corporate vested interests and the books consistently misrepresent the science.

One terrible example of the confounding of science with narrative arose in 1997, when Fred Seitz was collecting signatures on a petition refuting global warming. He did this by attracting attention to an article formatted as if it had been published in the prestigious scientific journal *Proceedings of the National Academy of Sciences.* The article carried a number of inaccurate and misleading claims about climate change as if they were science. Yet, despite its appearance, this article had never appeared in the journal or gone through the standard peer-review process. The journal felt obliged to hold a press conference to distance themselves from its claims and from Fred Seitz's petition. It was all sheer falsehood.

Science, by its very nature, is a discipline. It requires its originators, as Francis Bacon recognized so long ago, to pay systematic attention to detail and to develop a thoroughgoing rigor in reporting, if it is to yield its truths to us. There have always been bad scientists and poor science. When I was practicing as an immunologist, there was a great deal of work that was simply unreliable, because the experiments were poorly designed or badly reported. Yet the discipline was not threatened by this because scientific discernment was of the essence. You could ask the hard questions about a dubious study. You could point out the inconsistencies in the data. And this was possible because scientists met together and published together as a team of specialists determined to get at truth that we could all build on. That was the very essence of the pursuit. This movement in the United States regarding climate change, and a whole host of other important scientific positions, proves, sadly, that scientific truth is also susceptible to being undermined by influential people with an axe to grind. A closer look at how this happens reveals that it is a multi-faceted process.

The Anatomy of Subversion

The subversion of science, as it is currently being practiced, is not the sole responsibility of a few rogue individuals who are, for whatever deeper reason, seeking to protect 'freedom' as defined by neo-liberal economics. It is actually the product of a system that is weak at a number of points. The people I have described, who are twisting and contorting the public debates about scientific issues like climate change, work with other elements in society and depend on these other elements to co-operate, if they are to be effective.

First of all, there are the media. The media in Western society today do not carry simple news about science. They need their audiences and for that reason they will make sure that they both entertain audiences and play to their deepest commitments. They also want to entertain. Television producers are always looking for the angle that will enhance their reputation, provoke controversy and gain a wider audience. They need to do this. It is part of their job and a deep-level commitment of broadcasting in all its forms. And the problem is that dispassionate science is a teensy weensy bit boring. So they search for presenters who can make it interesting. The 'David Attenborough' or the 'Brian Cox'. They don't want to report some boring scientific conference. What they want to do is find a controversy. And there's the rub. Controversy is entertaining. Quite a lot of science is hard to make interesting to the public at large, but if you see people going head to head, with the adrenalin flying, then people can feel that and are captivated by the scrap, even if they are not bothered about the truth.

The media, of course, also have their own disciplines. One of these is to provide balance. In the world of narrative truth, this is very important. There is always more than one story about an event. There is always more than one perspective to give. And the truth is best served by the broadcasting of multiple viewpoints by our media, so that the general public can form their own commitment as to the truth of the matter in hand. This media discipline is formalized in outlets like the BBC, where providing balance is carefully weighed and reviewed. Yet it does not work for science. The media seems largely ignorant of the distinction between science and narrative in relation to the big public issues of today. If there is a scientific consensus that climate change is happening and that it is the result of human action, then the media need to report this, simply and incontrovertibly. After all, they know they would not get away with broadcasting flat-Earth theories, so why give time to those outside the relevant scientific discipline, who have silly 'opinions' about the science? This is a category mistake. The general public can and should debate the policy outcomes of scientific findings. It is in this arena that all 'opinions' must be heard. But they have no right to comment on the science, because they have not done, or rigorously studied, the experiments that established the position.

As a result of this confusion we have the ridiculous spectacle of the media feeling an obligation to provide 'balance' on scientific matters. They justify this partly by their accepted discipline toward narrative balance, but also because it makes for more interesting broadcasting. A stronger distinction between science and narrative could improve broadcasting practice considerably.

The media also carry their own bias at an editorial level. Serious analyses have now been made about the political bias and its effects on climate-change reporting. The 2013 IPCC report, mentioned above, was featured on much of the Western world's media, but there was a clear bias in the reporting. Right-wing media consistently gave more air time to climate-change sceptics.[1]

The second ingredient is the politicians. Politicians are complex creatures. They attempt to hold a host of competing positions in tension. They know they are formally accountable to the people, but they also meet a wide variety of stakeholders in society, some of whom are extremely influential. Keeping the stakeholders onside is a practiced art for the politician. When it comes to the influence of science on society, a politician will always be tempted to use science to their own ends and to ignore science that does not fit their purposes. They will also develop a set of trusted advisors to help them through this process.

One of the key attributes of the small cadre of influencers in the scientific debates of today is that they have power over the politicians. Nigel Lawson, of course, was the Chancellor of the Exchequer for many years and thereby will have gained respect at the highest places in the land. In the US, many of the antagonists regarding climate change won their spurs with the US government over defense issues. As a result no one would reasonably doubt their patriotism and, if they are suggesting a policy whereby serious and disruptive change can be avoided, then every politician would be tempted to keep consulting them.

So it is that politicians in a democracy are actually quite weak in terms of the potential for creating radical change, even if that change is necessary. There are few real leaders, in the sense of people who can stand up for what is right irrespective of the complex and diverse powers pressing around them. There are even fewer who can publicly change their mind. As became clear in Chapter Four, the nature of party politics is that politicians are signed up to deep-level commitments about what matters in the world and their policies tend to flow from that spring. If they believe in the 'free market' and 'minimal government' they find it very difficult to countenance serious intervention in society in response to scientific truth. No wonder they then look for trusted advisors who will tell them what they want to hear with an air of gravitas. It is a good way of keeping the peace. Apparently.

So these are the players in the tradition of scientific subversion, the small cadre of influencers, the media and the politicians. But there is one more that we have not yet discussed and it is arguably the most important of all. And that is us.

The Role of the General Public

First of all, perhaps we should consider how we are reacting to all this. As of January 2015, 88% of scientists 'believe' that climate change is due to human activity, but only 50% of the American public believe the same.[5] There is some, but much less, dissonance between scientists and the general public in the UK. The interesting thing about the opinions of the general public is that they now split along party-political lines, with Republicans in the US much more likely than democrats to be suspicious of the scientists. Back in the 70s polls about the environment did not reveal the same party-political split. 'Conservation' – as it was then known, was a right-wing type of thought and there seemed no reason for there to even be a party-political line on these things. Surely this could be a subject on which all agreed? But no. The party-political bias in the response to climate change has grown and is now very evident. In the UK, a similar political bias can be demonstrated with conservative-minded people being more likely to deny human-induced climate change. So where did this bias come from? Has the bias actually been formed by the media? That is certainly possible.

It is very clear that the type of reporting of climate change by right-wing media is sensational and designed to outrage its readers even as it plays to their deepest fears and worries. A good example of this is the work of Christopher Booker for the *Daily Telegraph*. Booker is not a scientist, but he is very quick to report anything that looks suspicious in the climate-change scene. There was a moment when he spotted a blog suggesting that some temperature readings in Latin America had been changed by the authorities. He exposed it in his column claiming the greatest scientific scandal of all time. I can imagine 'Outraged of Croydon' reading Booker's column and becoming furious, even as this article massaged his/her desire to write off climate change as a scam. Maybe the media works by saying outrageous things to confirm the prejudices of its readers? The temperature-reading story was, like so many others, just a damp squib. The scientists responded with a report showing how they continually monitored and adjusted global temperature readings for all sorts of reasons like changes in equipment/calibration errors, proximity to cities etc.[5] The whole story was actually baseless, but it may have had its effect on Outraged of Croydon and many others like him/her.

The overall dissonance between what scientists think and what the public maintain is actually found in a wide variety of subjects especially those involving safety, e.g. food grown with pesticides, genetically-modified foods or nuclear energy. In each case it is apparent that the public do not trust science or scientists.

It may appear from all this that the public are the helpless playthings of the great powers. Part of our weakness may be because we have accepted the description of ourselves as passive consumers and expect life to be done for us. The tragedy is that we may be colluding with the media and the politicians in the suppression of scientific truth and the neutering of our response to it.

The media and the politicians both want us to listen to them. In fact they are desperate for us to listen to them. In their bid to entertain us there are few lengths to which the media will not go.

Every potentially tedious element of a show must be enlivened with a new angle, which will entertain. Science must be mediated by personality. All simple truth wrapped up in story. Such is the way of the media. Another way of saying this is that our attention spans have become shorter and our entertainment choices greater, so the character of broadcasting has to follow the great spiral towards trivia.

Nevertheless the urge to communicate and learn remains. Some of us, at least, are offended by being called consumers and are ready to become citizens. We are ready to try to weigh public matters seriously for the good of all, but are confused as to how to make that intention clear.

Politicians, in their turn, would like to treat us like citizens, but feel obliged to listen to the manipulators. These 'advisors', as they are called, see the world in terms of messages that win votes. To be in their presence is to feel dirty because they treat people, as Kant would have described, as a means to an end, not an end in themselves. The advisors talk continually about keeping people onside in order to retain power. As a result the potential for the politician to take a stand for future, rather than present, good is highly problematic. They might say they would do this if they could, but the public's voting patterns are what must really concern them.

Yet there is another way. We could require the politicians to act responsibly for the future. They, in turn, could require the media to report seriously and responsibly. But so far we have failed in this great task. Instead we have become passive recipients of culture rather than shapers of culture. In the words of the title of this book, the time has come when the people must 'Rise Up' if we are to take on the challenges that we now face as human beings. And I don't mean by this phrase that we need to take part in some sort of insurrection. I mean we need to rise up in the sense of being citizens with a full responsibility for our collective life. We need to rise up in the sense of being genuinely imaginative about the future rather than letting others decide the future for us. And we need to rise up in the sense of challenging the dominant influences around us that are holding us in this passive, childlike mode of being.

Examining our Commitments

The subversion of science, as here described, turns around two distinct elements. The first is a failure to distinguish the distinctive roles of science and narrative in society. This could be remedied quite simply by reinvesting public conversation with new disciplines that recognized this distinction. So scientific congresses designed to get at scientific truth would be free to examine, probe, question and work towards scientific truth without the subversive glare of others. Meanwhile the whole policy area would be recognized as narrative, and public debate fostered about the best way to deal with the scientific truths with which we are being presented. Here there is room for difference. Indeed here there should be difference. It is also a place for imagination, leadership, appeal and hope. Making the distinction between science and narrative clear would be a good first step.

The second element of this problem turns around unexamined commitments within us all. I have described here how the powerful influencers in the US have come from a deeply concerned commitment about defense of that country and how this coincides with a commitment to minimal government and free market. In some cases these commitments will take the form of fundamentalist 'Castles of the Mind' (see Chapter Four) whereby commitments are taken on with extraordinary power as a means of simplifying the very complex world in which we live.

Similarly the media have taken on commitments related to their ratings performance and the need to entertain, which are subverting many of their more serious presentations. Some are consciously propagating right-wing views, telling false stories about wrongdoing to inflame their loyal readers. And we ourselves as the general public are committed to what? We certainly have things we care about, but so often they are submerged and we are tossed about on a sea of pleasure-seeking and dubious thoughts about freedom. Maybe it is time for us to examine our inner lives and release the power that comes from finding harmony within. This will be the subject of Chapter Ten.

One great step forward for society at large would be to deal better with narrative. In previous chapters I have tried to describe how important narratives are to our public life and how they express our deepest commitments. I would like to take this opportunity to describe more fully how a good society engages with narrative. It seems to me that the role of any leader in the world is to make an appeal to the society around them by constructing a narrative that takes into account any material evidence, e.g. the science, and imagines how human beings could and should react in this moment. Such a story necessarily comes from an empathic base. It will draw out how people might feel about the challenge they are faced with. It will appeal to them about a path forward that will produce a good result. It will be laden with emotion, will attempt to elicit hope and, if told well, may convince the people.

The people on the other hand will weigh what is said, not in a dispassionate way, but with the whole of their being, assessing facts, but also bringing their own empathy and emotion into their response. They will construct a narrative in response about what should be done. Other leaders may say different things and evoke different responses. People may meet and confer and hear conflicting opinions. So the process of narrative engagement goes on. At its best this process will engage all the people in conversation.

I was personally heartened by the recent debate about Scottish Independence, not so much by the result, as by the fact that the whole community appeared to have a very 'grown-up' conversation. I mean by this that the people were taking the issue seriously, hearing different views, coming to a decision. This is real politics. And, to my view, it is also the way we should be deciding the future of our planet in response to climate change.

As we hear the stories of others during such a process, we grow in our own empathic understanding. We appreciate aspects of the problem and its potential impact on others across the world. We learn together as a community. This is why a rich narrative culture is the key ingredient of a good society. Fundamentalists of all sorts are characterized by an inability to hear anything that challenges their narrative. They will behave badly in any serious conversation or attempt to dominate it, simply because they must shut out contrary views. A culture of true citizens is at heart a culture that listens well to different narratives.

A few years ago I set up a charity known as Agora. It was to be modelled on the center of a Greek city, where all sorts of people met and talked. It was the marketplace; the local authority met there. The whole of public life could be found there. So, the charity Agora was about drawing people into public conversation. It has since developed into something of a hub for social innovation. We have pioneered various ways of getting people together around food-growing, helped to form a local currency, and are now forming a new type of food co-operative. In all these things the very act of doing something new encourages conversation about society and how it works. I have come to believe very strongly that the true health of a society is not found in its Gross Domestic Product or its working hours or any of the conventional measures. The true health of a society is found in the quality of its public conversation.

Why Is This Chapter Important?

Science gives us a certain sort of truth, which we have to take account of in our public life. The deliberate undermining of science, described in this chapter, strikes at the heart of our civilization and could spell its doom. We need to be very clear about the different roles of science and narrative in a good society, listening to the established truths of science and debating the policy alternatives by means of narrative engagement. Unless we can distinguish these two forms of truth-seeking and their different methods, we will be continually prone to importing unexamined, false, or even malevolent commitments into the judgements of our public life.

Chapter Six

Short-Sighted Economics

In a Nutshell

The world of economics suffers from a similar confusion between science and narrative as does the struggle for truth regarding climate change, except that, with economics, the boot is on the other foot. Economics should properly be considered a narrative discipline, but we have pretended it is a science. This has resulted in a formulaic interpretation of economics as a discipline that rules out of consideration all sorts of vital human concerns like human justice or the long-term well-being of the planet. Economic formulae, expressed in neo-liberal economic theory, are supposed to yield maximum happiness for all, but have actually become a cloak for unexamined, self-interested commitments by powerful people and organizations, who determine our future. We need to open ourselves to a holistic view of economics that engages with our full range of human sensibilities and creativity.

Picking Up the Threads of the Past

Economic exchange has taken place within and between cultures for as long as human societies have existed. A city like Venice made market behavior a founding principle of its life.[1] Its churches and public places were decorated with gold, while economic calculations dominated the thought life of its citizens. Its market place, the Rialto, was designated a sacred site as a sign of its value to the city. Yet the story of Venice is not a good one. Such devotion to monetary gain proved corrupting. Famous for the creation of the first casino, the society eventually disintegrated through endemic, destructive gambling.

The UK also has a rich history of economic development, most notably in the formation of guilds, societies, joint stock companies and a national money system. Yet a profound change occurred in the eighteenth century, in the period following the Enlightenment. This was centered on an attempt to give economic activity a more strictly rational basis. Adam Smith, Thomas Hobbes and utilitarians like John Stuart Mill attempted to theorize about economics, trying to understand how economic prosperity could be assured, or how 'the greatest good for the greatest number' could be achieved etc. Their intention was to do something like science, in the sense of deriving an economic system built on logic, so that human progress in this area could be assured. The problem was that they were dealing with people, who were inevitably diverse and who could be described in a variety of ways. So how could you produce a theoretical scheme that had all the attributes of a science yet was based on human beings? The answer they came up with was the concept of an 'average' person. They basically said that people on average behave in a certain way and, if you can work out what that is, then an economy, which necessarily deals with people in large numbers, can be designed to work from a rational basis. Each of these founders of modern economics described the situation slightly differently, but each placed a series of fundamental commitments about people and their motivations at the heart of their theory. Adam Smith is famous for building on a concept of self-interest in market activities.[2] Thomas Hobbes conceived of society in terms of competing individuals, whose activities were moderated by the state.[3] John Stuart Mill conceived of 'Homo Economicus', a stereotypical person who would behave in predictable ways in economic life, so that he could build his 'system of logic'.[4]

This has produced what we now know as the dominant neo-liberal form of economics. It functions as a *formula* for the promulgation of human society that eliminates the need for any subtlety in culture. Far from a rich storytelling culture, there is a commitment in neo-liberal economics that the only proper way of conceiving society is as an amorphous collection of self-interested individuals working rationally to maximize their personal 'utility', however they perceive that. This formula is considered to be all that is required to ensure our progress as a society and claims to override all other human considerations. If you have a formula for progress, you do not need *stories* that negotiate what progress looks like. There is no need and no ability to consider the future in a responsible manner. You can simply tear into the rainforests, drill in the Arctic, quarry in Africa, until it is all gone and you are surrounded by your own mess. If you empty proper debate about success into measures of Gross Domestic Product, you can create extraordinary anomalies, like it being better to pay someone to look after a child than to encourage the child's own parent to do it. Likewise, if you perceive society as a mass of competing individuals, then you have no language with which to express the value of civil society and all the activities that traditional societies do outside of market exchanges. You relegate stories about such activities to the 'private' sphere and imply they are of little real value.[5] Similarly, you empty the political process from the richness of exchange of diverse stories, into a technical debate about implementation of the formula. All these things are currently happening around us and there is little real opposition to it. During the first decade of the new millennium all the dominant political parties in the UK were in thrall to the economic 'reality', as it was being promulgated by the financial institutions of London. In Europe it is still financial technicians that call the tune when a country like Greece is in trouble. We are generating a new super-elite of financial technicians who reward themselves with massive salaries and increasingly live in a social bubble of their own making. This is profoundly dangerous for the culture of UK, Europe and all those who are now aping

'Western' culture.

The Deep Power behind the System

The pseudo-scientific construction of economics has a darker side. There have been those who have recognized its shortcomings for decades. For example, a classic study by Herman Daly and Peter Cobb in the 1980s showed how economics needed to incorporate other measures into its thinking if it were genuinely to work 'for the common good'.[6] This has been followed by many other serious analysts making a similar critique. But their work has fallen on deaf ears and little has changed. The reason for this is that the formulae of economics have created a new class of super-powerful people and corporations who now control human society. These people will not relinquish their power in the face of some 'academic' study. They see themselves as the saviors of a benighted world, whose very power and riches will ultimately help all those in poverty.

There are many who have shown that such 'trickle-down' theory is not working and will not work. The idea that the neo-liberal economic theories will automatically provide for a just and prosperous world are simply untenable. Social polarization has been increasing in both the UK and US over the last forty years. In the 1970s the highest 1% of earners in the UK received 5 % of all the income distributed in the country. Today the same 1% of earners receive a full 15% of all the income distributed. The International Monetary Fund research, on which these figures are based, shows that such shifts of income towards the rich actually diminish the overall level of economic growth.[7] Others like Richard Wilkinson and Thomas Piketty challenge the neo-liberal model. Wilkinson, in his acclaimed book *The Spirit* Level,[8] reviews the sense of well-being in various countries as reported in a raft of indicators. He finds that the countries with a greater equality of income and opportunity have a notably higher sense of well-being. Piketty[9] suggests there is a flaw in neo-liberalism and that, understood correctly, the existing system predicts this recorded rise in wealth of the richest people. He says that unless it is addressed by the state, it could cause a democratic collapse. Both these critiques are substantial and serious, yet they have had demonstrably little effect. Because those who hold power do not need to listen. They have escaped all real accountability.

The reality of this deep power behind the pseudo-science is made manifest in two vital areas. The first are the multinational corporations, whose own economies often dwarf those of the smaller nations and who are funded by a set of shareholders, whose interests are solely in financial return. They are led by the oil and gas industry, which provides most of the power on which the whole economic edifice depends. Their phenomenal power in the marketplace is only matched by their ruthless behavior in regard to the natural world. They are masters of public relations with enormous marketing budgets, well able to create campaigns that insist that they are on track to move 'Beyond Petroleum' or whatever their PR team deem will enhance their image. Yet the reality is that these corporations have to respond to the short-term interests of their shareholders and they have little power to buck this trend.

Ben van Beurden, Chief Executive of Shell, speaking at the annual general meeting in 2015, said that he was unable to invest more heavily in renewables or carbon capture and storage because they would not produce the same high financial returns that his shareholders were used to. He admitted the short-termism of his shareholder thinking. Recognizing the importance of these new technologies for our global future, he confided to the annual meeting that "I have great difficulty to have shareholders focus on the quarter after next."[10]

Other multinational conglomerates control important sectors of the world economy. They are also funded by shareholders, whose only interest is short-term return on investment, so they cannot take any sort of moral lead. In the food sector there is Cargill and ADM, which together control around 75% of the world's grain trade. They are also primary players in the fertilizer market. International treaties, which are purportedly simply implementing 'free trade', actually create the conditions whereby the trade dominance of these huge multinationals is consolidated. And that, you could say, is their 'real' purpose. It is the perpetuation of deep, pervasive power that cannot be checked according to a more holistic human narrative of purpose.

The Transatlantic Trade and Investment Partnership (TTIP) is the latest manifestation of this approach. Yet the dangers are made clear by recent information that has come to light about 'endocrine-disrupting chemicals' (EDCs) in pesticides. These chemicals change our body systems when present at remarkably low doses in our water and the EU was preparing criteria to manage their levels in pesticides, but it took just one trade delegation from the US visiting European officials under the guise of TTIP for the whole matter of EDCs to be dropped. This illustrates the sinister purposes of trade deals such as TTIP. They are really about stopping countries exercising their own concerns about products and their social or environmental impact.[11]

The crucial factor in this discussion about the power of multinationals is the role of shareholders. If they, en masse, can only attend to their self-interest, then the world is stuffed with them in the driving seat. This calls attention to the structure of our financial systems, as they have grown in response to our neo-liberal economic orthodoxy.

The Power of Money

Money systems grow in every society as systems of trust, but our own money has recently had a spectacular history. Banks have taken the power to issue money to the world. Issuing money used to be the province of governments, but now banks are allowed to do it and are responsible for around 95% of the money in circulation. The way they issue money is by making a loan. The monetary value of a loan does not exist anywhere before the bank manager approves the loan and is not removed from any ledger when it is loaned. It is simply created by the loan. The money is, of course, lent at interest and such interest gives the bank power over the future. The twentieth century saw a succession of banking crises, with a tussle between governments and banks for power, but this was finally resolved in the 1970s when Ronald Reagan took the US right out of the conventional gold-standard system and made the US dollar the de facto reserve currency for the world. This liberated the banks and prompted the massive rise in bank speculation on futures and other markets in the world.[12]

The global financial system almost crashed in 2008/9. Many years of speculative growth, borrowing to invest and pretending that 'greed is good' finally reached their apogee. Yet it was fascinating to note how little change resulted from this crisis. The UK government propped up some banks, but made few changes to the way that the system worked. The US did similar. It was like they were dealing with some precious idol which they dared not offend. And perhaps that is an apt description. Ever since the Thatcher era, the UK has focused on London as the powerhouse of the British economy. Manufacturing has largely been outsourced and we are now an economy based on the riches that accrue from financial speculation. It is a position of phenomenal weakness and, if nothing changes, is sure to bring our ruin. This is why the divide between 'the city' and the rest of the country has grown so dramatically. Some people in London simply do not have the time to spend all their wealth. They go on sprees of lavish consumption at weekends. Agents for elite cars park up outside the financial houses at bonus time. And they 'achieve' what they do by gambling with other people's money.[13] I wish this were an exaggeration, but I am assured that it is not. We are told that this is how the system must work. Free trade must have its way. The economic formula must remain in place.

Mark Carney, Governor of the Bank of England, said this in May 2014 to a conference in London:

> *Just as any revolution eats its children, unchecked market fundamentalism can devour the social capital essential for the long term dynamism of capitalism itself... All ideologies are prone to extremes. Capitalism loses its sense of moderation when the belief in the power of the market enters the realm of faith. In the decades prior to [the financial crisis of 2008] such radicalism came to dominate economic ideas and became a pattern of social behavior.*[14]

He went on to urge a reform of the financial sector before it destroyed itself.

These words also reflect the earlier critique of Max Weber, one of the founders of sociology as we now understand it. He demonstrated the importance of an ethic derived from Protestantism in the foundations of capitalism.[15] The inner motivation towards honesty and integrity, towards hard work, frugality, saving and a sense of calling produced a brake on conspicuous consumption that was, in his view, crucial to the development and maintenance of the prosperous society envisaged by capitalism. Weber realized that once this was undone, there would be no countervailing power to the great machine of economic growth that we had unleashed. People would exist as if in an 'iron cage' of their own making.[15]

Other commentators have demonstrated the power of this mechanism over the Earth itself.[16] It is tearing into the very fabric of the Earth as you read this book. It seems that nothing can stop it. Climate change looks to be proceeding inexorably. Mass extinction of species is already occurring. Food prices will rise.

People will become desperate and then all hell will break loose.

So it is time for the rest of us to rise up and live differently. The last section of the book will explain what this means in more detail, but for now, I would like to draw attention to some overall principles that can counter the debilitating weakness of our current position.

The Social Challenges of Scale

Globalization was the watchword of the twentieth century. With the unravelling of all local laws and practice in the economic world, the stage was set for companies to grow to true global dominance and so to dictate the behavior of markets by the sheer power of their trading volumes. Yet something was being lost in this process. A global economy tends to lose touch with real people and their situations and with real nature and its complexity. The person buying clothes in the UK really could not know how they were being made, what sort of sweat shop was producing this 'oh-so-cheap' garment. So a raft of organizations set out to tell them. Campaigning groups, like the World Development Movement, were complemented by the rise of ethical brands like FairTrade to try to redress the social imbalance buried in the realities of global trade. And they have done a good job. But is it good enough? Can this approach solve the problem? Many developing-world countries remain in massive debt. They are forced to adjust their economies to the dictates of the IMF and WTO with an emphasis on creating cheap exports to pay down debt rather than giving attention to their own infrastructure and the long-term resilience of their economy. Something structural has to change.

I believe that the heart of a new approach to economics can be found in one simple idea. That is that wherever we are in the world, we should attempt to localize those parts of the economy that can be localized. This sort of shift has the potential to engender really positive social and political effects and provide a new resilience, especially to poorer economies. It is also an absolute necessity in terms of action on climate change.

The current system fails on several counts. Firstly, our transportation strategies, both for people and goods, are based on a falsely cheap price of oil. The price of oil is so low, because the price does not account for the impacts of oil extraction and burning on the natural world. If we had a mechanism for pricing in the real costs of oil then we would recognize that our way of life had to change, and that change would be in the direction of localization.

Some will say, in response to the problem of oil dependency, that we can replace fossil fuels with other sources of energy and, of course, we can and should do this. But those who crunch the numbers recognize that there is *also* no alternative but to cut down on our energy expenditure as a society. Renewable energy will not resource the sort of lifestyles and transport of goods that we have come to expect and demand. We need to cut down on our energy use and that can be achieved, in part, by greater localization. This might, at first, seem like an unpleasant constraint on our lives, but a more localized existence actually has powerful, positive social repercussions.

A more local economy will act as a spur to community life. The mere act of localization will encourage people to talk about life together and what they are seeking to achieve. It will allow the sort of engagement of narratives and embrace of feelings that the first section of this book commended. I can see local co-operatives arising to compete with global brands. I can see a mass of ordinary people finding strength and initiative to begin new grassroots enterprises. I can also see different ways of life emerging across the world and being tested in a way that our formulaic and manipulative economy currently cannot permit. There is a terrible uniformity now imposing itself across the world. Often dominated by American ways of thinking and acting, it is the inevitable outworking of a globalized economy. It is no coincidence, for example, that high streets are now the same right across the world. Whatever big city you go to, you will find the same brands. How sad is that for humanity? What a dreadful indictment of our creativity? It is like the big brands are colonizing the whole world. We should not put up with this.

The difficulty in making a transition to a more local economy remains the current legal structure. Economic law demands that one does not privilege the local in transactions. Naomi Klein tells the story of Ontario, Canada, where there was an ambitious plan to replace their dependency on coal with solar power and to do so by teaming up with a Silfab, a global leader in solar technology. Their contract with Silfab was also to state the requirement that materials should be sourced locally and workers should be hired locally. For a time the deal looked on and the region was looked to as a great example of innovation. Then the World Trade Organization (WTO) stepped in and declared their 'buy-local, hire-local' provisions to be illegal. These provisions were deemed to distort the market and act against those outside the area who would bid. This sort of legal intervention is a serious limitation on the renewal of local economies. Naomi Klein comments that the real distortion in the market, which WTO does not recognize, is the massive effective subsidy that is given to fossil-fuel companies because they are not charged for their pollution of the atmosphere.[17]

Money with Values

When the history of Western Society is written, one of the great tragic stories that will be told will be the gradual corruption of its money supply. How did it come to be that the UK economy was in the hands of boys in red braces, snorting cocaine and demanding a Lambourghini?[18] When did we decide to allow this? When did our Queen or our government formally relinquish control over the economy? Of course we didn't do any of these things. We just found ourselves caught up in a process whereby those who managed the money supply for us grew to control it. The controllers then began to develop ever-more-sophisticated instruments with which to assess value. They expanded the use of these instruments in a global market working at lightning speed, lost touch with the reality of these instruments and crashed the system.

The ordinary person in the street was caught up with the falsely booming economy in terms of rising house prices. All that money that the banks were creating had to go somewhere, and much of it was lent out as mortgages to finance house-buying. House prices went up and we all thought we were getting richer. But actually we were more in debt. The overall amount owed by householders on their mortgages grew like topsy.[19] Then it all began to fall apart, beginning with the collapse of the strangely named US mortgage-lenders Fannie Mae and Freddy Mac; the whole financial system tottered in 2008. The politicians huffed and puffed as if they were in charge, but knew that they could not allow the banks to fail. They ran round with sticking plasters and resisted any substantial reformation of the system. The result of all this is the rather terrible fear of debt now stalking governments and resulting in austerity programs, while everyone is well aware that financial collapse remains possible at any moment.

One rather different approach that gets round some of the legal hurdles is to recognize our potential to create 'money with values'. By this I mean money that is, by its very nature, working to develop a different kind of system. There are many ways such 'conditioned money' can be created. We have developed a system here in Bristol using a local currency known as the Bristol Pound, which can only be redeemed by locally owned businesses. We are also about to launch a business-to-business credit model based on a non-interest-bearing loan. Such initiatives can provide new ways of approaching the free-trade laws that proscribe against privileging the local economy. They can also foster the diversity of the local economy and strengthen community life. I will describe these initiatives further in Part III of the book.

Why Is This Chapter Important?

This chapter has tried to show that for all our ingenuity as human beings, we have found ourselves in a hopeless mess in terms of economics. There are crucial flaws in our current method:

1. the intention to mimic science led early economic theorists to look for building blocks on which their theories could be constructed. As atoms were understood as the building blocks of science, they drew an 'atomistic' picture of society, with the human being perceived as *an individual in a competitive relationship with others.* In fact human beings have developed as social creatures. As such they are not only *able* to work cooperatively with others, but their 'success', however it is defined, is actually predicated on working effectively in cooperation.
2. the mechanics of exchange were conceived of as centering on *self-interest,* which is no doubt a factor in most people's lives, but is not an adequate summary of their relationships, even in the marketplace.

The result of this dread simplification of theory around economics has been an extraordinary perversion of human culture. The supposed scientific basis for economics has allowed it to develop as the primary means by which human society should be coordinated and, as a result, it has undermined our culture, subverted proper political process and is doing violence to our relationship with the Earth.

This chapter has argued that a process of re-localization can reinvigorate community life and help people engage with their feelings and so recreate powerful narratives about the world they live in. Such re-localization will also allow our communities to cut energy use in response to climate change and allow the creation of new financial systems.

Chapter Seven

Political Failure

In a Nutshell

It is clear that our current political system is failing to engage with the big questions that are facing humanity. We need to make major strategic decisions now if we are to avoid a climate catastrophe, yet the last two elections in the UK have scarcely mentioned the subject. This is stark evidence of political incompetence. This chapter explores why Western governments are unable to address these issues. It concludes that we need to rethink the scale of our democratic structures. Nations are a relatively recent invention and are now failing. With the majority of the world now living in cities, it seems that city regions may have the necessary political competence to negotiate our future and that a parliament of cities could work on a global scale.

The Origins of Politics

Our sense of politics arises from our lives as social animals. How does it feel to be a member of this community? Am I safe to go out? Am I alright if I bump into strangers? Can I rely on the established ways of this community so that I can trust it enough to be creative, to play my part and know that others will play theirs? These are the grassroots questions of politics. If they can be answered in the affirmative then life feels good.

One way of understanding political regimes is to see their goal as creatively upholding the social order, while allowing appropriate challenge and change. Social orders are an extraordinary example of trust that arises from a deep sense of shared values. Many aspects of these social orders can be negotiated without words, as shown by primate colonies and discussed in Chapter Two. I say this because there has been a tendency in more recent times to see politics as a formula. We have adopted the word 'democracy' as a signifier of all that is good in politics and tend to act as if this democratic good can simply be imposed on struggling communities and nations by means of a formulaic process. The truth is rather more complicated.

Social orders are actually very fragile and derive from a complex interaction of values, beliefs, institutions, history, identity and much more. It can seem that a new social order is just around the corner, yet the journey may prove impossible. The Arab Spring of 2010 seemed, at first, like the sort of popular uprising that could only do good. Western governments looked for ways to support the burgeoning and apparently 'democratic' movement that was sweeping Arab nations. Yet the process of change has become a nightmare. Challenge has turned to civil war and protest to a dangerous chaos. One lesson from this process might be to learn the value of social orders and recognize that attempts to remove even a bad leadership, as in Saddam Hussein's Iraq, Gadaffi's Libya or Assad's Syria, can have disastrous consequences. Once people can no longer go about in safety, once they cannot trust their common institutions, then political space is made for the new and often harsher power that can restore public order. It is the same in the playground as in the prison, in the public school as in the wilderness. The absence of the creative trust of a peaceable social order makes space for the bully, the formation of new 'orders' in the form of gangs, or harsh religion like Islamic State. This is why good politics is an art and simple, assumed formulas an illusion.

The Source of Political Failure

The UK may have the most long-running and stable democracy in the world. Although everyone traces democratic institutions to Athens, the UK experience has particular strengths, unfolding, as it did, over a period of many centuries and attaining strength through the complexity of the trust that developed over that time. Some would trace the originating source of the UK experience to the inevitable struggle of a monarch and the nobles in a relatively defined territory. The monarch needs to find capacity for some great public project, like a vainglorious war, and has to negotiate with the nobles in the regions for soldiers and money. So the tension is set up between the king and 'others', who were firstly the nobles, later the landowners, then business people and finally all adult people, so that eventually a plethora of voices could be heard in public space. Such so-called 'Whig histories' trace a line of developing accountability for the monarch from a moment like the Magna Carta through to today's parliamentary democracy.

Many things are assumed in these accounts of political development. Firstly the geography of the British Isles tends towards certain sorts of political solutions. When national boundaries are largely set by the sea, then they acquire a permanent and given status. The Danes, the French and others may have invaded us. They may have invaded, settled and intermarried, but the whole set of people inhabiting the islands were sure to create their own political identity in the end. As a result the people of the UK easily 'assume' the idea of nationhood as the heart of their politics.[1]

Meanwhile, in the rest of Europe, the story of political development takes on a very different hue. In much of Europe the nation state remained elusive until very recently. Europe may not have developed the concept of nationhood, as we now have it, until after their great European wars and the Peace of Westphalia in 1648. It can be argued, in fact, that the whole idea of nationhood is a British export, most notoriously in Africa and the Middle East where we worked with other colonizers like the French to draw lines in the sand, like the currently contentious Sykes-Picot line demarcating Iraq, Libya, Syria etc.[2]

Once the general idea of democracy had been established and academics had justified it, then the space was cleared for democracy to be imposed by revolution. Tom Paine, for example, played a founding role in the democracies of both France and the US. Such revolutions have a special fervor about them and tend to uphold the elements of their initial constitution with extraordinary zeal. It is as if, without a monarchy, the people have a psychological need to affirm the constitution, because they know that the whole social order is dependent on it. France and the United States demonstrate this zeal for their constitutions.

By contrast, when democracy develops gradually, a people can easily take it for granted. A constitution may never formally be adopted and many citizens may never ask the deeper questions about what it means or even how it works. Such is the case in the UK today. Few ordinary members of our society really 'understand' democracy in its practical outworking or have thought through its implications. They may complain about it, but few embrace a common narrative about how it should work. Some confuse 'representative' with 'direct' democracy and think that a 'true' democracy must mean everyone voting on everything. Others are dismissive about the whole business. The tragedy of such 'assumed' democracy is that it is not subject to the proper scrutiny of its citizens and our politicians can fail us badly without provoking a response from the people. The story of the Iraq War is a case in point. The Chilcot enquiry, which was supposed to get to the bottom of how this travesty of democracy occurred, has still not reported at the time of writing (Jan 2016), more than twelve years after the events. How can this have been allowed to happen? The citizens have allowed it to.

One little-discussed political theme is the matter of power or competence to act. Any political regime needs to be able to engage with the decisions that affect the lives of its citizens. The history of nations indicates that they have found their most obvious focus in terms of national defense and the prosecution of war. Even today in the UK, it is parades of soldiers and the remembrance of those who have died in war that bring out the most powerful nationalist feelings. National defense is of vital importance and yet defense is only one aspect of a good political regime. There are other powers that are essential for good government.

The globalization of our financial and economic structures, as described in the last chapter, has meant that national governments are less and less powerful over the life of their nations. More and more of the major strategic decisions of their life are being made by global systems over which they have no control. This was dramatically demonstrated in the financial crisis of 2008. Banks are now massive global corporations and the financial crisis was necessarily global in extent. Everyone, including some of the bank employees, was shocked at the extent of corruption in banking practices revealed by the crisis and yet the UK politicians felt they had no alternative but to prop the banks up. Since 2008 Western governments have struggled with the decisions they made in those moments of the financial crisis. The massive public debt caused by this debacle bears down upon them and many have decided to impose regimes of 'austerity' to try to balance the public debt. Yet despite this, they rarely criticize the banks or dare to undertake their substantial reform. The UK has put itself in a peculiarly invidious position by hosting global banking services in London, thus making our economy terribly susceptible to bank failures. As a result we dare not enact any substantial program of bank reform. This betrays a failure of political nerve and competence. It reveals the unpalatable truth that the financial and economic systems of the world have largely escaped political control and are now essentially running the show as puppeteers of national governments.

The seriousness of this situation is compounded by the fact that our global financial and economic systems operate in an environment defined by self-interest from which all values of a more co-operative nature are defined out of consideration. So it is that behind every publicly listed company in the world is a set of shareholders, and their advisors/brokers, who are set to work only to further the self-interest of the individual investor. This, they will contend, is not out of any malevolence of the individuals involved. It is enshrined in the system. It is the way they must act. As a result, it is no exaggeration to say that we have handed world power over to a set of self-interested individuals who are 'gaming' the system. This is the political outcome of our 'free-market', formulaic economics, which now dominate our public life. The situation needs to be compared with how a good representative democracy needs to function.

Necessary Politics

Politics is a manifestation of power and accountability. The social systems that make up human society are ineradicably complex and require various political structures for good governance. Part of this variety needs to be expressed in a multiplicity of *levels* of political organization from family, to neighborhood, city, region, nation and world. The whole idea that 'nations' are the default position around which all else must dance is a dangerous distraction from creative and responsive thinking in this area. This is particularly important because the very shape of human societies and their communication potential has been dramatically changing over the last few decades.

Any political regime must have a set of competencies; that is, areas over which they have genuine authority. These need to be clearly set out and agreed at the next highest level. For example, a city might say to a neighborhood, "You can manage the parks in your area and you can raise a local tax to do this." Or a nation might say to a city, "You can manage education in your city and you can raise an education element, as part of your local tax, to do this." In each of these and other cases a good democratic political structure should ensure that major strategic decisions are put to the people and that the outworking of these decisions is tested by some form of 'parliamentary' process.

For example, at a national level, we might have a general election where different strategic approaches to governance are put to the people in the form of party manifestos or we might debate devolution as in Scotland in 2014 or our inclusion in the European Union, as promised shortly. The Scottish experience was an extraordinary example of good politics, with a massive proportion of the people actively engaged in an intelligent debate on an issue of vital importance to their future. The referendum on membership of the EU may prove similar.

A central part of these campaigning strategies should be to articulate the deep values that are at stake. These are what inform the narrative they present to the people. In some situations the values are clear and implicit in the debate. In Scottish devolution, for example, the whole process clearly hinged around national identities, common loyalty and how that worked out pragmatically. The European debate may likewise turn on the extent to which UK citizens *feel* themselves to be part of Europe as well as the practical considerations for trade, benefits, migration and other contentious issues. The difficulty with UK politics in recent years is that the only values that have been evident are those that relate to the free market. Different parties have slightly different strategic approaches to the same issues, but they all use the same language to describe the problems, waxing lyrical about 'jobs', 'growth' and 'GDP' without referring to any deeper sense of purpose and values behind their policies. One party may talk more about 'small government' but this is nothing more than an outworking of their free-market commitments. Another might talk about public investment, but their reference point remains GDP. In my view this sort of values-empty thinking is nothing short of a tragedy that results in public boredom with the whole process. The primary reason for the election of Jeremy Corbyn as leader of the Labour Party, the growth of UKIP and the Green Party's own surge in membership could be that they are all willing to articulate values other than free-market pragmatism. People feel what they are saying both positively and negatively and there is an impassioned public debate.

Having set up a general election or referendum on major strategic direction or reform, a good political structure must then set up a parliamentary type of process to scrutinize the implementation of the decisions made. Here there must be real engagement with elected representatives of opposing views. This debate should be communicated to the people via a plethora of media and an attempt should be made by the government in power to forge the best solution.

These are some very simple ideas that underlie democracy. The problem with our situation today is that the structural forms we have inherited are seriously ineffective and need reform.

Towards a Better Political Settlement

Over the last fifty years there has been a gradual diminution of power at what is called the 'local', as opposed to the national, level in the UK. This has been a deliberate policy of national governments, which cities and regions have found little ability to counteract. It arose from some startling stories of inadequate government in the 1960s and 70s, in a few cities, where a small cabal of local leaders was governing against the public interest. Much of the problem in these situations appears to have arisen through a failure of public scrutiny. Attempts have since been made to reform local government with city leaders, or mayors, who form an executive cabinet of publicly accountable figures. Despite these changes, national governments remain very reluctant to devolve power to cities. Likewise cities have also largely failed to devolve power to neighborhoods. At a conference on mayoral governance in my home city of Bristol,

UK, representatives from the national civil service admitted quite frankly that the only way for cities to obtain the power they needed from national government was for them to 'take it'. Quite how this should be done remains somewhat mysterious.

So how should political systems be designed? Is there a right level for a political structure? How should we think about devolution or should the national system always call the tune? It may be that there are three key criteria for discernment about good political practice.

Firstly, the citizens must identify, at the level of their deepest feelings and concerns, with the *level* of political government envisaged. This is a matter that is truly 'beyond words' and therefore only partly amenable to analysis. Yet when a government dreams up a new 'region' and gives it a name, or when a local government creates wards along boundaries no one recognizes, then the system is sure to fail. Part of the issue here is that all our local identities have been dissolving in the face of globalization. People simply do not feel a sense of belonging to place in the way that we used to. We travel all over the world for pleasure. We shop using the internet by calling in goods from all over the world. Meanwhile, our local area can seem little more than a convenience, with an identity to be laid aside whenever the next housing opportunity arises. Such is the situation of many people in the UK in 2015. Yet politics has, until now, always been constructed around our dwelling places as the ultimate sign of our identity and the root of our political structures. Perhaps this is no longer appropriate? The sheer weakness of local identity in a globalized, market-driven society may be forcing us to consider other political structures. How to work in tune with our deepest feelings and concerns is a mystery we need to solve.

The second characteristic of good political practice is that politicians should have real power and the public should understand what power they have. This sounds very straightforward but anyone who has practiced politics knows how much their hands are tied. The public regularly expect them to deliver on changes that are outside of their brief or tied up in legislation they cannot change. Explaining this to the person seeking redress can be very difficult. Often the answer to the question "What can be done?" is "Not much," which does not leave the political representative in good odor.

The third characteristic has to do with passion and integrity. I firmly believe in the power of articulating one's own values and commitments on a political stage, not as some ideological and unchanging dogma, but as the real concerns of a flesh-and-blood person trying to live well in the world. Those who empty themselves into procedural replies and established policy statements may defuse tension, but will ultimately bore.

So how does this work out today with our changing patterns of relationship and the diminishing grip of our inherited systems? My sense is that the most important change might be moving our prime political focus from the nation to the city region. I realize this idea may come as a shock to the reader, but the idea of a city state has a long pedigree, much longer and more illustrious than that of the nation. Athens itself was, of course, a city state. So were Sparta, Rome, Florence etc. Much of the world's history has actually focused on the history of cities. Cities have a natural competence that arises from them as centers of trade and innovation. They tend to have diverse cultures represented within them, which makes them liberal in disposition, and they have a strong and natural interest in their own well-being. In addition their relatively small geographical area allows for strong communication between citizens, with the potential for face-to-face meetings, artistic and musical contributions to culture and the potential for great festivals that unite and celebrate the common life. City regions also develop natural concerns for the people of their surrounding region and the ongoing sustainability of the whole area.

This natural predisposition to city-state politics has become even more significant in recent years in terms of the mass movement of people from rural areas to cities across the globe. This migration has been so strong that more than 50% of the world's population now lives in a town or a city. In a 'developed' country like the UK, this figure rises to 80%. The attraction of cities comes largely from the economic and cultural opportunities that cities present, as contrasted with stable, but often harsher lives in rural areas.

So could it be that we should re-imagine our political structures so that the focus turns around city regions rather than nations? Such a system would have some important things going for it. Firstly, city regions right across the globe face a common problem. It is how to conceive of the city and its hinterland as a sustainable entity in terms of its relationship with the planet. In the previous chapter I mooted the idea of re-localization as a key to economic reform. It may also be a key to political reform. If cities are given the power to manage their affairs and to tax their citizens, then they could realistically engage with the difficulties of sustaining our relationship with the planet. They could design and finance their own transport systems, their own energy systems, even their own currency and financial systems. Many people are already scanning the world for good examples of cities and their regions developing sustainable practices. It could be that discovering how to develop sustainable city regions is the way to crack the big problem of climate change.

The focus on the city region might also be part of a greater movement towards the localization of economic life that will necessarily strengthen community relations. The impulse towards localization will stimulate imagination in cities as people consider what they could do for themselves as a community. As a result localization will give real and measurable purpose to community activities, calling into being all sorts of new associations and initiatives, building 'social capital' and encouraging deeper and more meaningful community relations.[3] This could, in turn, result in further devolution of power and more engaged politics at a neighborhood level.

The movement towards participatory budgeting is a fascinating example of this type of neighborhood politics. In this activity a local authority typically sets aside a pot of money for community projects, then invites members of the community to put forward their ideas for what might be done to improve their community. Public events are held to encourage project submission and then the authority does some feasibility work on the ideas so as to create a shortlist of fundable and achievable projects. Finally, the citizens of the area are invited to vote for the projects of their choice. Every citizen can vote and those chosen receive funding and support in the execution of the project. There are extant examples of this type of activity from many places across the world, but few in the UK.

A Global Politics Based on City Regions

Nations have found it very difficult to co-operate across borders with other nations. They learn to trade with 'other' nations, but otherwise, quite naturally, take defensive, self-interested positions in regard to other countries. The reasons for this may lie in the long history of nations as organizations of defense and the prosecution of war. They have simply not been constructed around values that encourage co-operation with the one perceived as 'other'. This should properly be conceived as an empathic failure (see Chapter Two) and may have been understandable in the past, when most inhabitants were unaware of the wider world, but it is now inexcusable. This property means that nations are wholly ill-equipped for the next phase in human history, which must include co-operation on a scale greater than humanity has previously achieved.

The UK has particular difficulties in this area, as shown by our most recent concerns about being members of the European

Union. The European Union was born out of the two great wars of the twentieth century, with a resolve by its members to forge something that would be more than a trading bloc. The EU would actively provide a sense of mutual solidarity for those who had previously made war with one another. This is the spirit that lies behind the EU. Its members see it as of value in and of itself and as a sign of co-belonging. Yet the UK does not share this feeling. We do not *feel* European as others do. Hence the Tory Party can get elected on the promise to offer a way out.

It is very hard to see how the world can respond to the challenges of climate change and our generally destructive approach to the planet. Those with a commitment to 'small government' get the heebie-jeebies about climate change because they identify response with the establishment of a new dictatorial world government that will override nations, their identities and values and prevent free trade, which they see as the only route to world prosperity. This reaction is a great pity. There are many ways of conceiving co-operation across the world without invoking a dictatorial form of world government. And there are many ways of re-imagining trade so that it does not destroy the Earth.

It could be that the way through this sort of conundrum is a renewed focus on the city region as the focus of political co-operation. Nations are frightened by cooperation with other nations in a way that cities are not frightened by co-operation with other cities. Similarly, national politics often aligns itself with deep psychological impulses, like the desire for order and suspicion towards change, but city politics offers a refreshing new possibility.

The history of city mayors shows a very interesting trend. City mayors tend to be more independent in their thinking than politicians of national parties. Even if they come out of a political party, a figure like Boris Johnson presents as an independent thinker, willing to cock a snook at party ideology and act pragmatically for the city he believes in. This new loyalty, namely loyalty to the city, its people and its culture becomes the focus for a new political identity, whose good transcends traditional party allegiance. It is interesting how many city mayors across the world present as creative, fun-loving people, who dare to be different. As mentioned earlier, in my own city of Bristol, we elected our first political mayor three years ago. George Ferguson is an architect of some standing in the city, previously outside the political system, and a person who is clearly daring and imaginative. He is unafraid to talk about the city being fun-loving and encourages all those engaged with festivals and feel-good events. He obviously makes enemies as well as friends and the election next year will test the balance of that, but I think no one would doubt the fact that he has made city politics more accessible to the citizens of the city.[4] The prime difficulty with the arrangement appears to be ensuring the elected councilors in the city also play a meaningful role in governance.

I believe that a global parliament of mayors might provide a new avenue for the development of truly creative global governance. It is very unlikely that nations are going to be able to respond to the challenges we face. The great international conference on climate change in Paris 2015 resulted in some commitments, which were framed boldly, but are very unlikely to be enacted, because the participating nations will prove unable or unwilling to deliver on them. Yet a parliament of mayors governing city regions could set targets for themselves, could measure them and could spur one another on in their achievements. There are real possibilities here.[5]

To establish a new global

Why Is This Chapter Important?

This chapter has described how our current political arrangements in the Western world, and in the UK in particular, are no longer fit for purpose. The nation state should not be treated as the default political arrangement even though the political history of the UK has led us to believe that this is the case. Nations have defined themselves by war and defense against aggressors. By contrast, world history has demonstrated the creative power of cities and their leaders, with cities demonstrating their potential as places of multicultural tolerance and economic innovation.

With the mass movement of people towards cities today, the global problem of climate change can be broken down into the problem of creating sustainable city regions. If cities can apply themselves to the problems of being genuinely sustainable and learn from one another's innovations in this area, then the problem of our human pollution and destruction of the Earth's natural systems can be cracked.

Chapter Eight

Human-Centered Religion

In a Nutshell

Religions have formed an important part of societies in the past, but they are facing a unique set of challenges today. This chapter argues that the religions of the world formed their identities at a time when the Earth was perceived as a vast, limitless, wild space and encounters with nature were marked by dread and awe. The key problem for religions was therefore perceived in terms of human-to-human relationships and the ordering and transforming of societies. The whole idea that the Earth and its resources are finite is new to religion. Some religious groups also continue to struggle with science and the new powers it has given to us. This chapter outlines how religion has to change if it is to contribute well to life on our planet in the twenty-first century.

The Importance of Religion to Society

Many people today dismiss religion as something of the past, which has been superseded. I intend to challenge that view by offering a more subtle approach to understanding the impact of religion on society than has been the conventional fare of the sociologists of religion.[1] I suggest that we can best understand the true function of religion by considering the types of imagination that a particular religion evokes and the particular set of empathies that it enhances.

The best way into this might be to consider the believer in prayer, or whatever they count as prayer, and ask the question, "What aspects of their imagination are engaged in this process?"

and "What parts of the world of their feelings come into play?" As I hope to show in the paragraphs that follow, these are the great determinants of the impact of religion on the society in which believers are placed. Different religions stimulate the imagination in very particular ways and this has a profound impact on the believer and their behavior. There is a clear distinction to be made, for example, between the more poetic imagination of mystical religions as compared with the sequential and very rational imagination of Protestant Christianity. There are also some very different empathic sensitivities and values that result from this imaginative process. The settled values of a religious person are often a life commitment. They will sacrifice for these things. They will give of their energy, their time and their money to express their concerns. Such is the way of faith and that is why these things are important to understand.

The Changing Scenes

Spiritual people of all ages and times have headed out for the wild space in order to encounter God. This is where their imagination was formed. Moses went out into the desert, where he heard the call to lead his people out of slavery. Elijah went to a cave to find God when he was burnt out. Jesus found solace in the wilderness lands. The Buddha likewise sought enlightenment in the wild space. For all these people, nature was an unchanging reality, an assured space where all the struggles and changes of human society could be put in perspective and new inspiration found.

Yet today our perception of the world is very different. Some might say that with the rise of science and technology and the progressive urbanization of the human species, we have despoiled nature. Gerard Manley Hopkins wrote eloquently of our current condition:

Generations have trod, have trod, have trod;
And all is seared with trade; bleared, smeared with toil;
And wears man's smudge...[2]

Bill McKibben likewise wrote a book provocatively entitled *The End of Nature*.[3] His thesis was that the whole idea of the wild, unexplored and uncharted space was no more. Human beings have been everywhere and left their mark. The corollary of this, for people of faith, is that nature does not have the same capacity to inspire or connect with the divine. For many people, nature is now a problem. It is something to study, dissect, analyze. And few of us fall down in awe.

In times past, the sheer evocation of the wildness of nature would be enough to inspire faith. Consider the words of Job as he mocks the puny abilities of human beings in the face of wild nature:

> *Can you draw out the crocodile with a fishhook or press down his tongue with a cord? Can you put a rope in his nose or pierce his jaw with a hook? Will he make many supplications to you? Will he speak to you in soft words?... Will you play with him as with a bird or will you put him on a leash for your maidens?... Lay hands on him. Think of the battle. You will never do it again.*[4]

Yet science has forced a profound change of perspective upon us. We feel more powerful through science. We sense that we have conquered wild nature. The difficulty is that we have never learnt to really care for it. I will argue in this chapter that the true object of religion in this day has to be to help us care about our relationship with the natural world and note, with some dismay, that our existing established religions appear unable to embrace this task.

Different Complementary Perspectives

The perceived opposition of religion to science has been told through a few primary stories like the history of Galileo Galilei, for example, and illustrates that the Church, in his time, was indeed confused about science. I think that some measure of confusion at the outset of a new wave of scientific endeavor in the world is very understandable, but it does not mean that religion and science need to remain at loggerheads forever. Part of the argument with Galileo was about the way that the ancient scriptures should be interpreted. For example, should a psalm that declares, "Thou hast set the earth on its foundations, so that it should never be shaken," be interpreted poetically or literally? The answer to that question was not obvious to the Church at that time, but I think it is now quite clear. It should be read poetically.

On the basis of the analysis given in the first four chapters of this book, it is obvious that there are different ways of paying attention to the world and genuinely different and complementary benefits from each process. Science pays a certain focused attention to the material world and creates a view of the world as an assemblage of its constituent parts. By contrast narrative-based knowledge, and in here I count religion, takes a holistic view of the world, engaging with our feelings, our motivations, our values and our commitments. The Roman Catholic Church has developed a teaching around this called 'Non-Overlapping Magisterial Authority' which was championed by no less a biologist than Stephen Jay Gould. This contends that science and religion exist as two separate spheres of knowledge each with their own proper authority. A similar but different type of description derives from McGilchrist's work on the left and right hemispheres, whereby the scientific way of seeing the world is largely related to the left hemisphere, while religion arises from the holistic, empathic and relational world of the right hemisphere (see Chapter One). I argue in this chapter that the true object of religion in this day has to be to help us care about our relationship with the natural world and note, with some dismay, that our existing established religions appear unable to embrace this task.

Part of the problem is that many people remain unclear about the relationship between science and faith and so create unnecessary tension between the two. Here are two examples from the contemporary scene.

In a workshop developed by Joanna Macy, called the Council of All Beings,[5] participants are encouraged to go out into the natural world and to stay there until they are aware of a plant or a creature, which they are being called to represent. They attempt to feel the world as the plant or creature feels it and then return to the group, where a 'council' is held, bringing together the feelings of the natural world around. During the council, the participants speak out, often very evocatively, for their plant or creature they are representing. Different language will be used about the experience in a Council of All Beings. Some participants will talk of being possessed by the spirit of something. Others will speak simply in terms of imaginative empathy. In some ways the process is reminiscent of the prophetic tradition where the prophet speaks up for the poorest and most marginalized members of the community. My point in regard to this discussion is that there is nothing in this process that in any way conflicts with knowledge accrued through science. It is bringing a different perspective to the world, one that embodies feelings and generates a moral position with regard to nature.

By contrast, there are some claims in the Green movement that are a direct challenge to science and should be examined on that basis. For example, the Steiner movement, in relationship to biodynamics, has developed a series of 'preparations', for which there are recipes. The preparations, when spread across a field, are claimed to generate a superabundant fertility. This language is confrontational in regard to science, because it purports to define conditions on which a repeatable experiment can be based, and a scientist would be right to ask to see the data on which these claims are founded. Similar observations can be made about other aspects of biodynamics, like the way that viticulture is performed in relation to the lunar cycle and its effect on the quality of the resulting wine. Such ideas can, and should, always be tested by experiment.[6]

These are both modern examples, where distinctions should properly be made between scientific and religious claims to truth. The established religions of our day present a whole set of similar problems which, if they could be untangled, would create a new space for religion to contribute to the world. And yet there are many sides to this. One is the relative silence of the Church on matters to do with the natural world. Creation is very much on the official Church agenda. It believes in a Creator. Many Christian denominations recite the Nicene Creed each week, declaring:

We believe in one God,
the Father, the Almighty,
maker of heaven and earth,
of all that is,
seen and unseen.[7]

Yet the amount of time and energy devoted to the subject of Creation in the churches is pitiful. Why is this? As one who has practiced as an Anglican priest for more than thirty years and as a research scientist for more than ten years before that, I have some personal understanding of this subject.[8] I have come to the conclusion that one of the most serious reasons for churches ignoring the subject of creation in their teaching today is that the ministers do not feel confident in this area. This is because the debates about evolution and the origin of the universe have tended to pit religion against science and ministers have felt the need for specialist help to tackle the subject at all. Scientists, such as John Polkinghorne for example, have given an enormous number of talks to religious audiences, offering ways of understanding religious language that do not conflict with mainstream cosmological science. Other scientists have done the same for the parallel evolution debate. Yet, the upshot of all this activity is that 'Creation' has become a specialist subject, which you must invite a person with scientific credentials to speak on. The ordinary parish priest simply does not feel equipped to comment.

Another way of describing what has happened here is that a genuine religious insight about the 'givenness' of the natural world, with a resultant moral imperative to 'care' for the natural world, has got lost down the rabbit hole of a highly specialized debate about origins. The real problem that religion can and should address is our ongoing relationship with the natural world.

This is a very serious situation, but it pales into insignificance in comparison with an even more profound difficulty in this area.

The Deep Problem with Religion

I guess if you are going to be critical then the place to start is with your own people and faith. I will move from there to consider the reasons why *all* the established religions appear to be failing to address the challenges of living in a better harmony with the Earth.

For the last ten years or so I have worked with the confident assumption that the Christian religion could be brought to the realization that creation lies at the very heart of its belief system and, from there, it could take an effective public role in helping us all feel a proper concern for our relationship with the Earth and its creatures. My work included penning a book entitled *The*

Dream that Inspired the Bible, attempting to show that the hope that lay behind Judeo-Christianity involved harmony with the natural world. It also involved attempting to start a movement called EarthAbbey to create a spiritual home for people with a Christian and Earth interest.

I was frustrated from the start with the amount of engagement that this topic elicited. Meetings would be sparsely attended. Book sales low. And I thought, "Is this me? Am I doing something wrong?" Then I began to meet with other people in the same field. Organizations like *A Rocha, EcoCongregation* or *Christian Ecology Link* (now *Green Christian*) have been much more successful in the field than my own initiatives, but, still, I found a sense of frustration among the leaders of these movements. There was so much more that could be done, so few, in the overall picture, who were responding. My friends had to be all upbeat about what they were achieving in public, of course, because people don't join a failing movement. But behind the scenes there were some deep concerns.

If you considered the pronouncements of religious leaders in this area, you might think that all is well, but this is an illusion. We have had a succession of Christian leaders speaking out on climate change. At the *Summit of Consciences for the Climate* in summer 2015, for example, many of the world's spiritual leaders were gathered and they made suitable speeches about their concern for the Earth. For example the Papal representative, Cardinal Peter Turkson, who co-wrote the most recent Papal Encyclical *Laudato Si* on climate change, said, "At stake now is the wellbeing of the earth, our common hope. What we need is care..."[9] They were fine words and such sentiments were echoed around the conference. Yet there can be a vast difference between the leaders' pronouncements and the extent of concern expressed by the religion in practice. These leaders are acutely aware of their public standing and most recognize that part of their job is to articulate the concerns of the faith in relation to society. They will 'speak out' on any and every 'issue' that they are presented with. By contrast, the people on the ground, in parishes and congregations, are remarkably disengaged with these issues. They have their own lives and, led by their ministers, they often act quite independently of their overall leaders. In fact their whole religious worldview is often shaped more by their inherited tradition than by the pronouncements of their current leaders. And there are problems with the inherited Christian tradition.

For me, the amazing feature of Christian belief is the intimate sense of God. Jesus taught, in a way that few others ever have, that God could be known as 'Father'. This gave Christian faith an extraordinarily personal heart. God was to be known as a 'person', as a parent, and could be approached as such. It was a breakthrough in religion and continues to be inspiring to me and many others. As Christianity developed, the inter-personal status of its formulations was underlined with the concept of the Trinity, of three persons in one, a great mystery, which is repeated by today's worshippers in both detail and regularity as they recite the historic creeds of the Church.

Yet there was a difficulty with this approach. The clarity with which God was now understood in personal terms led to the faith being thought of in terms of human–human relationships. God was 'like' a human being. Human beings were made in the 'image' of God. These were two sides of the same coin. The outcome of this system of thought is the general assumption that God is *primarily concerned with human beings*. So Christianity becomes locked into a world of human–human relationships and any concern for other creatures or the well-being of the natural world becomes an added extra to the 'real' work of the gospel as it is perceived by the ordinary member. Inter-church bodies might make decrees about the 'marks of mission' that include the 'integrity of creation' and suchlike, but this has little impact on the faith of the worshipper. Every week they proclaim God as

Father. Every week they pray for the very human society around them. When they confess their sins, they are likely to use words like, "We have sinned against God and against our neighbor." The word 'neighbor' *could* be expanded by some imaginative theologian to include the natural world, but this is not what it presents to the ordinary worshipper. They are being subliminally taught that sins are against humans, not against the natural world.

Religious art gives an insight into the perception of the faith down the centuries. Popes would commission art from talented artists. The artists themselves were highly creative people, but their patrons defined their subject matter and deliberately used their art as a way to market the faith to the masses. When religious art is assessed in this way, as the popular message, even the advertising, of the Church, then its human-centeredness becomes immediately apparent. Tour around the Vatican Museum and what you see is a mass of bodies.[10] This faith is *about* human beings. Enter the Sistine Chapel and all around you are depictions of bodies, while above you a very human-looking God stretches out a finger to touch humanity. The human-centeredness of the faith could not be more obvious.

The Protestant Reformation in Europe, of course, provided a profound change in the nature of religious imagination. A new and much more rational approach to faith came to dominate the minds of the religious. This was, at one level, much more consistent with the thought forms of science. The Protestant faith could be 'proved'. The faith was reformed so as to base itself on the scripture and the sermon as a way of removing many of the apparent excesses of the Church. Yet, at the same time, the very rationality of the faith meant that it trod on the toes of science and would come into an irreconcilable conflict with it.

Unfortunately the Protestant Reformation did little or nothing to enhance religious imagination in regard to nature. The faith remained human-centered. It may no longer depict this in the form of art, but the center of the faith revolved around very human sensibilities. God, as compassionate Father, was the object of religious devotion and Protestant churches, as well as those in the Catholic tradition, would express their compassion for other human beings in a rich variety of ways. Believers would give their lives to serving the poor or campaigning for social transformation, inspired by what they perceived to be the concerns of their very human conception of God.

So we have a problem. The genius of the Christian faith may prove to be its downfall. Describing God in such intimate human terms may implicitly deny God's concern for the rest of creation. This is a big worry to me, and as far as I am aware, it has not been noticed by others. We have, of course, recently become clear about the gender issues implied by the 'Our Father' prayer and some steps have been taken to address that, but I am not sure that we are as aware of the limitations of a 'personalist' description of God in the sense of what it does to the rest of creation.

Prophetic Failure

The type of human-centeredness exhibited by Christianity is obviously different from any human-centeredness in Judaism and Islam. The religious places of Judaism and Islam do not portray hundreds of bodies, nor is the divine being obviously personalized to the same degree as in Christianity. Yet these and other religions may still be human-centered in a deeper sense. Let me explain how this is so.

Every religion, in its very nature, takes root in a culture and has the capacity both to affirm that culture and to critique it. Religious history suggests that a religion tends to work particularly strongly for social transformation in the early stages of its life, then this is followed by an establishment period, where it becomes part of the institutions of the society and challenge is more muted. Nevertheless religions tend to retain some ability to critique the societies in which they are placed even after they have become established. They can provide a surprising and interesting challenge to governments at important moments. Such 'prophetic' activity is a vital part of democratic balance.

For a religion to be able to challenge the society within which it is established, there needs to be a sense of dissonance between the moral sensitivities of the religion and the ways of that society. The religious component of the slave trade was one such example. Stories of the treatment of slaves were clearly in breach of the Christian traditions around love for strangers. So the campaigners had a handle to work on. But today there is a problem. Human society is faced with a new *type* of crisis which requires a new type of moral sensitivity. In the long period of human history, when all our religions were established, there was no recognition of our potential to despoil the Earth and its creatures on a world-wide scale. The Earth was seen as vast and nature as wild. The sheer magnitude of the Earth created awe in the religious. Yet today we need to recognize the finite nature of the Earth and our immense capacity to destroy. It is like we have to invent a new sin, which religions have not previously articulated: a sin against the Earth and against the whole balance of life in which we participate.

In one sense our religions cannot be blamed for this failure. Their moral sensitivities were formed in a different age. They will naturally guard and perpetuate their historical moral tradition. What it does mean is that we need a new expression of religion for this age.

The Impact of Urbanization on Religion

This problem of prophetic failure in relation to the Earth is compounded by urbanization. Despite having no feel for the finitude of the Earth or our capacity to despoil it, most religious traditions did arise in direct contact with nature and their first adherents were people who knew something of the natural world and felt for it. It is no wonder, therefore, that religious dreams of peace included harmony with nature.[11] Fast-forward to the modern world and we are faced with increasing urbanization. It seems that people love cities and they are now flocking to them right across the world. The multicultural nature of a city necessarily interrogates religious people. It asks hard questions of them, which a more isolated rural life might not. It forces them to interact to some degree with others who think and believe differently. These things can be seen as a strength or a weakness depending on your position, but there is no doubt that the very existence of cities challenges established religions.

One aspect of this challenge is divorce from the natural world. People who live in cities tend to be dramatically less aware of nature. Most of their time is spent in a world of bricks and asphalt. They might carve out little spots for gardens or allotments, but most of them can pass most of their time in complete ignorance of the natural world and any call it might have on their concerns. They might think about nature, even go to talks about nature, but they miss the daily awareness of the Earth and its creatures that our rural ancestors enjoyed.

There are some people who have been aware of the general impact of modernization on our minds. Most of them are poets. Few of them are explicitly religious. John Clare, for example, mourned the enclosures, as they were destroying the world of his day.

Now all laid waste by Desolation's hand,
Whose cursed weapons level half the land.
Oh! who could see my dear green willows fall,
What feeling heart, but dropt a tear for all?
Accursed Wealth! o'er-bounding human laws,
Of every evil thou remain'st the cause:[12]

The great movement towards urbanization has also had a direct impact on religion. The harvest festivals, rogation tides and other great festivals of rural communities have found only a very muted expression in the cities. As a result the concerns of farmers are a lost voice to the religions. The human being, now rebranded as consumer, finds food at the supermarket, blithely unaware of how those humans and animals involved in the food chain have been treated. And those of faith often do not think of these things. They are of little concern compared to the great work which is to evangelize and win faith adherents. It is interesting to see how faith communities easily adopt a club-like mentality in the cities. Their job becomes to perpetuate and grow the club with little real thought about their impact, or lack of impact, on the wider society. Maybe it is because the city is such a smorgasbord of projects and people. It is easiest to stick with what we understand. Most religions of the past measured their success in terms of their impact on society. It could be that they were responsible for an aspect of society like poor relief, or that they were upholding the law in some way, or even campaigning for change. Yet they were seriously engaged with the wider society. Today's religious organizations tend to be concerned for their own. They establish their own projects. They look to their own finances and buildings and believe in their own particular way. And part of this change may be a response to the complexity of urbanization.

Emphasizing Values

It is a little-recognized truth that most religious adherents are attracted by the feelings and motivations that are generated by the faith far more than any theory of faith that they are presented with. So, for example, a Christian worshipper may appreciate the feelings of history in a building, as it connects them with people and society of the past. They may love the stories of Christ, particularly where Jesus demonstrates compassion for others.

They may feel compassion themselves as they enter into prayers for those in various kinds of need. They may enjoy the presence of children in the service, as the young people elicit caring feelings from those around them. Or they may love singing songs that evoke community feeling. This is the reality of faith for many people.

The slightly strange thing is that ministerial training does not concern itself with these things. Instead the putative minister is inducted into a whole range of arcane knowledge about the history and philosophy of their faith. As a result the minister becomes an 'expert' to the congregation, the one who can handle any difficult question, while other members simply enjoy the experience. This strange bifurcation cements the authority system in the church and is emphasized in a weekly sermon experience, whereby the minister is charged with speaking to the congregation.

The modern world finds it difficult to relate to the dogmatism of traditional religion and I think it is unlikely that the complex formularies of traditional religions will find broad acceptance again in our city-dominated world. This suggests that a whole new approach is required if religions are to play a positive role in society, and I suggest that the heart of this new approach could be a focus on values. Values like compassion, kindness, honesty and integrity are an assumed part of most religious faiths, but I would argue that they need to be brought to center stage. There are several reasons for this.

Firstly it is in relation to values that the real motivations of ordinary worshippers lie. Anything that changes their *experience* of church is suspect, because their feelings are the key to their faith. Every minister recognizes how hard it is to change anything, but often fails to name the feelings and values that lie at the heart of the other person's worry. If all the faith adherents, including the ministers, learnt to talk about their feelings and values, they would manage difference much better.

Secondly, values conversations will help faith groups relate to the wider world. There is evidence that we are beginning to recognize the importance of values in our public life again. The Prime Minister has even taken to talking about British values as a good thing that we should all embrace. It is sad then that the only such value he seems clear about is 'tolerance', which has no real energy in it.

Energy is the real heart of a vibrant value system. If we are clear about the values we embrace, then we will find a true passion to live by. I shall dwell further on this in the closing chapters of this book, but for now I want to simply note that feelings and values tap into the well-spring of energy within us. We will work, and give, and join with others to promote the things we care about.

A New Approach to the Sacred

I believe that there may be a whole new approach to the sacred available to us today that has not been available to earlier generations.

Religious people are used to calling particular things and places 'sacred'. The word might be applied to the area around an altar, for example, which only a priest may enter. The definition of the word is associated with that which is 'set apart', 'inviolable', 'to be approached with care' and 'of ultimate value'. The interesting thing about such phrases is that they chiefly describe an *attitude* towards something or an *experience* of something and they do not seek to actually define the divine.

Hold that thought for a moment and journey with me to the thinking of various non-religious people today who seem to be approaching very similar ideas from a secular standpoint. I have been aware for some years now that many of the most radical grassroots environmental activists promote attitudes of profound awe in response to the natural world. They may not naturally use the word 'sacred' in this context, but their experience of nature certainly speaks of 'approaching with care' and rendering 'ultimate value' to the natural world. I sense such attitudes among allotment holders, who are pursuing food-growing from the love of it and who feel something special about simply being at their allotment. Some will say, quite simply and unaffectedly, "This is my church."

I have heard how many people are deeply inspired by site tours around Permaculture projects. Permaculture is a way of thinking about the world that stresses harmony with the natural world and permaculture projects are often led by inspiring communities, who are pioneering new ways of living.[13] Trips around such Permaculture projects can be life-changing and are reminiscent of the way pilgrims used to tour around shrines. There is also a strong awareness of the potential for our relationship with the natural world to have healing effects, as it enhances well-being and builds community. As I understand it, these and many other people associated with renewal of our relationship with the Earth are all looking for a new language. I wonder if a new approach to 'the sacred' might provide this, because it reflects the attitude that they feel deep inside, yet makes no explicit claims to define a particular system of belief.

By ascribing the attitudes and experience of the sacred to the natural world, we are able to assign ultimate value to our relationship with the Earth and its creatures. At the same time we can acknowledge that the proper place of human being is not one who stands apart and judges the Earth, but one who participates with grace and sensitivity.

Such an attitude might be shared by both religious and non-religious people alike and contribute to the change of perspective that we so desperately need at this time. The latest papal encyclical, *Laudato Si*, declares that the present ecological crisis requires a "profound interior conversion,"[14] even among the religious. The problem is that the very conservative exposition of the tradition which is found in the papal encyclical provides no real resource for such a conversion. It is clear that we need something more than traditional religious fare.

Wendell Berry has been something of a pioneer in Christian thinking towards the environment. He is quite clear that Christianity has failed in its task. He asks, "How can modern Christianity have so solemnly folded its hands while so much of the work of God was and is being destroyed?" Berry uses quite clear and unequivocal language of the sacred in relationship to the Earth. He says, "We are holy creatures living among other holy creatures in a world that is holy."[15]

Such an attitude to the sacred could also make a bridge with science. For many years now science and religion appear to have been locked in a hostile embrace. The roots of this hostility are deep, formed out of the very ancient roots of religion and the reforming zeal of science, but the philosophical untangling of science and religion as different types of knowledge (see Chapters Three and Four), presents us with a new opportunity. This approach to the sacred could be fully scientific, appreciating the detailed understanding and knowledge of the creatures with whom we share the planet and the dynamics of the Earth system in which they live, while also evoking the most profound respect in a religious and holistic sense of the balances, movements and intricacies which sustain life. One scientific writer who has impressed me in this area is David Suzuki, whose writings contain an exquisite feeling for the Earth.[16]

In Chapter Eleven I will propose several ways in which acknowledgement of the Earth and its creatures as sacred can be expressed by all people today, and provide a way in which together we can 'Rise Up' to meet the challenges of this generation.

Why Is This Chapter Important?

This chapter has outlined how religions across the world are failing to respond to the environmental imperatives that we are faced with. It has underlined that there is no necessary antagonism between religion and science because they each represent different and complementary perspectives on life.

The particular strength of religion is found in its ability to evoke and maintain feelings and values about matters of concern, but this is compromised, at least in Christianity, by a deep-seated human-centeredness in the way that it thinks about God and God's concern.

The chapter advocates focusing on values as the expression of feelings common to religious and non-religious people and the adoption of a new approach to 'the sacred', which is about treating the Earth and all its living systems with profound respect, according it an ultimate value and seeing ourselves as participants in the great dance of life.

Part III

Rise Up

I could feel the muscles in my wings.
They were strong and supple.
I rose on a current of air
to a place
of great energy
and calm

Chapter Nine

Facing the Great Challenge

In a Nutshell

The first two parts of the book have described pervasive weaknesses in human society and how these are playing out in some of our major institutions. In this last section I want to explore how these weaknesses can be addressed. I also want to recognize that it is absolutely vital that we do address these faults, because we are about to face the greatest challenge that has ever confronted human beings, namely the urgent need to live in a better harmony with the Earth and all that lives. We need to pay attention to our relationship with nature, our own inner lives, the strength of our communities, our ability to work together in a common cause and the role of leadership.

This chapter sets out a high-level strategic approach towards a sustainable life, based on city regions as the key unit of strategic development. It suggests a single core strategy, namely localization of all aspects of life that can be localized, as an outward means of addressing our needs at this time. Four key drivers of change are identified: finance, energy, food and governance. Interventions in each of these areas have the potential to catalyze change in the whole system and give citizens hope that there can be an alternative way of living in the world. Particular local examples of this process are given in the areas of finance and food.

The Great Challenge

This has not been a standard book about climate change, because, in my view, climate change is not a standard problem. I believe that climate change and, more generally, our whole relationship with nature, is the most serious problem to have ever confronted the human species. It is more serious than the ice ages, more serious than the plagues and more serious than the world wars. This is the big one. And it is unclear at the moment whether the human species can respond adequately to the challenge. By the end of this century it is very possible that the human species will be decimated; that is, reduced to less than one tenth of its current population. Whole cities could be lying in ruin, like Chernobyl, with nature re-establishing itself once more over the concrete. And the path to that moment could include the most terrible misery and destruction that the planet has ever seen.

This book has not set out the evidence for climate change. Others have done that with great rigour[1] and few have listened. This book will also not commend the trivial marketing methods that governments originally promoted in order to elicit individual behavior change, because this problem goes way deeper than anything an individual can remedy on their own. It is a collective problem, carried forward by institutions and corporate thinking that have their own logic and momentum. It is also an inner problem, to do with our perception of the world, what we expect of life and the way we work together. Previous chapters have reviewed our potential as human beings and the various traps we have fallen into that have brought us to this semi-paralyzed state at the very moment that we are faced with the ultimate challenge. And the big question is: can we rise to that challenge?

We have huffed and puffed about climate change for more than a decade now,[2] but little has happened. The public have actually become less sensitive to this issue than they were ten years ago.[3] Part of this hardening might be put down to a sense of impotence, as the massive nature of the challenge dawns and the individual feels no possible role in its resolution. Coupled with this is the politicization of the cause so that major public figures have come out and aligned climate skepticism with conservative politics and neo-liberal, small-state approaches to life that are non-interventionist at their corc. This plasticity of opinion on a scientific matter proves, as was reviewed in Chapter Four, that people are not as rational as they pretend to be.

Somewhere we have to find hope that we can deal with this problem.

A World of Sustainable Cities

The mass migration of people into cities is a global phenomenon that is unlikely to reverse unless total chaos ensues. More than 50% of the world now lives in a city and this proportion continues to increase. We love cities for their opportunity and diversity, for their pragmatism and creativity, for their hope and vitality. Yet this global demographic change towards living in cities presents us with a new possibility. We could see the city, or the 'city region' to be more precise, as the focus for global sustainability. In other words, if we could work out what a genuinely sustainable city looks like then we might have cracked the global problem. Cities could embark on a learning exercise, each exploring their own geographical and economic context, how sustainability might arise there and what they need to do by way of adaptation. The cities could learn from one another, taking examples of good practice from others, and so we could leverage our way across the world towards a sustainable way of life.

I would like to propose that the high-level strategic key to this process is a form of localization. This means that a city region looks to do for itself everything that it *can* do for itself. There are, of course, some aspects of life that are, in their very nature, so specialized that it would be daft for every city to do its own. This might include transport-vehicle manufacture, information technologies and the like. But there are many other systems that can and should be localized. In these cases localization does two things. It produces a dramatic decrease in carbon emissions and it strengthens human community life. The way that this might be achieved is through the employment of 'change drivers'. Interventions in these particular areas have the potential to act throughout many systems of the city, connecting people and communities in new ways and changing our deepest perceptions about life. I recognize four key change drivers:

1. Finance

Here in the UK we take sterling for granted. It is seen as *the* currency, and the idea that there could be any other currency is often laughed at. Yet it is clear that our currency system is part of what holds our current economic system in chains, that our banks and financial markets are now in such a powerful position in the world that they have become parasitic on the real economy of goods and services, drawing out its lifeblood while feeding an elite class of human beings.[4] It is also evident that our national politicians are powerless to address this. One way forward is to consider the role of local currencies and currency systems, not to replace national currencies, but to create alternative, localized systems of exchange that support the transition to a sustainable city. I will give a specific example of this below.

2. Energy

We have assumed that energy comes down power lines, which crisscross the nation forming a supply network that is assured and accountable to politicians. Yet, as with finance, a local supply can also make sense as a means of creating local efficiencies and resilience and complementing a national system. Few people realize the enormous losses in energy that are due to transmitting energy along power lines in the current AC system. George Monbiot, in his book *Heat*,[5] has detailed what can be achieved by local combined heat/power solutions, which involve power generation close to the places of supply, creating much greater efficiencies and, at the same time, drawing communities together in a common project.

3. Food

The food systems of a city offer an extraordinary example of the potential to reconnect citizens with nature. Everyone needs food and spends a significant proportion of their income on food, yet very few are currently aware of where it comes from. The dominant supermarket system insulates the 'consumer' from the ethics of production. We are simply unaware of the massive, industrialized farming operations that produce our food, the huge barns of chickens so fat they can't stand up, the mountain of wasted vegetables cast aside because they are the wrong shape, and the harvesters whose conditions are scarcely better than slavery. We don't recognize the plight of the supermarket suppliers as they incur enormous debts to meet a supermarket contract, only to have that contract threatened when the buyer casts around for efficiencies. These are the hidden elements of supermarket food-shopping. A veil has been drawn over them in our minds, so that all we ask is "Where can I shop most cheaply?" There are few more potent examples of our passivity today.

City regions could work systematically to increase the proportion of food they source from local producers. This would have the effect of bringing citizens in touch with local farmers and growers, making them more aware of the conditions under which their food was produced and also more sensitive to the seasons and the soil. Sourcing food locally could achieve phenomenal changes of perspective for the people of a city, helping us feel and value our connection with nature, not just as something pretty to be looked at, but for all its practical potential in providing food. The process would kick-start conversations about how food should be farmed, how farmers themselves protect or disrupt natural systems, what a fair price is for food, all questions that are currently beyond the ken of many. Some practical interventions in the food system will be described below.

4. Governance

National governance is now the default condition in the UK. Local authorities have been progressively emasculated for reasons explored in Chapter Five. It could be that the time has come for a serious attempt at devolution of power toward city regions. The whole concept of a sustainable city requires an administration that is created over an appropriate region and its empowerment to make the decisions necessary to transform its life.

In Bristol, UK we used to have an administration based on 'Avon' an area of around a million people extending around the city. In the 1970s, this was broken up into four areas, in what was described to me by a politician at the time as a 'Faustian pact' between the political parties. The Tories knew that they would rule the hinterland, while Labour anticipated its continuing dominance in the heart of the city. It was a tragic example of political interests that were not aligned with the best future for the city. A sign of this failure is our city's transport system, where various proposals have been put forward over the years, enormous design costs incurred, yet political feuding between the city and the hinterland authorities prevents really creative and innovative solutions.

In this context the UK government's recent proposals for devolution based on mayors that are accepted over the metropolitan region appears to be a step in the right direction.

A Case Study: The Bristol Pound

Most people in the UK think that money *is* sterling and *can be* no other. This is how a dominating system works. It tells you that there is no alternative. But maybe you can have an alternative currency? And one that is deliberately loaded with values, such as an intention to localize those parts of the economy that can be localized, to reinvigorate community life, and offer resilience against future shock. Back in 2009, a few of us gathered in Bristol around two questions. What would it be like if we made a local currency at scale, across a whole city? What would it be like if we had an electronic means of exchange as well as a paper currency? The Transition movement was booming across the world at that time and people were becoming aware that serious changes had to be made to the very structure of our economies. We had benefited here in the UK from some brave experiments. Totnes, Lewes, Brixton and Stroud had all launched a paper currency. They had proved that these systems could attract attention to local traders, but we wanted to explore the possibility of creating systemic change, truly benefiting the local economy by means of the local multiplier effect,[6] and for that there was a need to step up in scale. Our commitment was to launch a complementary currency across a whole city, with a means for both paper and electronic exchange.

Some people said we were brave; others, slightly mad. But the very small team grew from two, to three, four, six, then eight, as we put together careful and detailed plans for what we would do. Someone volunteered the idea that we could make payments by text, that the technology was all there and simple to execute and, to prove it, he built us a prototype. We burrowed into the detail and wrote a hundred pages of feasibility study. Our aim was to show people we could actually do this.

From the earliest phase we recognized that our crucial challenge was to attend to public confidence in the system. Some people were bound to say it was 'funny money', not quite real, and certainly not to be trusted. So, in response, we committed ourselves to backing every paper Bristol Pound with a pound sterling, which would be locked away in a trust account so that even if we went bankrupt, people's money would be safe. Similarly, we built confidence through a partnership with Bristol Credit Union, who agreed to manage our electronic accounts. Bristol Credit Union are regulated by what is now called our 'Financial Conduct Authority' and already managed accounts to a value of several million pounds.

The second challenge regarding public confidence concerned significance. We were aware that the public would quickly judge a local currency as to whether or not it would ever be significant in the life of the city. If they thought it would just be a gimmick, they would not participate and would just wait for it to melt away. So we went to the local authority and said, "We would like you to consider taking Bristol Pounds in payment for business rates." At first they were hesitant. How could they realistically change all their accounting systems for this thing that was likely to fail? Pressing on with the detail, we found a way that the public could pay their business rates in Bristol Pounds without causing a headache for the council. And so the deal was done. And our potential businesses knew that, if they had Bristol Pounds and found them difficult to spend with other member businesses, they could always pay their business rates. This proved particularly important in our trader recruitment.

So, we had some detailed plans that we were confident in, and our next task was to engage the city. Our paper currency needed designing. We had some great designers working *pro bono* for us. So they created a template and we put the major part of the design space out to the city as a community design competition. We received more than two-hundred entries from professional designers, keen amateurs, and many school children. Then an independent panel, drawn from different communities of the city, chose the winners.

The design competition attracted attention from the media, and from another unexpected source, the Bank of England, who wanted to enquire about our 'ambitious' scheme. As a result we found ourselves in Threadneedle Street, visiting with people from the Bank of England, the Treasury, the Financial Services Authority and the Financial Service Compensation Scheme all at the same time. The meeting was slightly nerve-wracking, from our point of view, but actually very constructive, as we cleared off many of the legal questions at one meeting and left with a renewed confidence.

Several years had passed by now. We had begun our explorations in 2009. It was now January 2012 and we faced the big decision. Are we ready to launch? It was big, because we hardly any money. Aside from a large donation from a trust for the purposes of developing our electronic software, we had almost nothing. We had just about enough to print the paper currency. By now we had a team of people who were somehow able to give enormous amounts of time with hardly any pay, but we realized that to say we were going to launch was to give a commitment we could not draw back from. We went for September 19th 2012.

As we drew closer to the launch day, the atmosphere in our office grew noticeably quieter and more intense. A trial of the text-payment system was under way. Exchange points were being developed across the city. The individuals and traders were signing up to the system. And it all had to be ready for the 19th.

The Corn Exchange lies in the heart of the historic part of Bristol. The short pillars known as the 'nails' in Corn Street are places of ancient exchange, where people swapped gold for goods, paying, as we say 'on the nail'. On the 19th of September at 12 noon the Lord Mayor, standing by one of the nails in all his fine regalia, held up a Bristol Pound and declared, "What will anyone give me in exchange for my Bristol Pound?" And a local trader stood forward bringing a loaf of bread and said, "I will give you this loaf for your Bristol Pound." Then the cameras whirred and we had a party.

The media frenzy that started that day has not really stopped.

We had coverage not only from our BBC, but from media providers from Europe and around the world. I know some people in Bristol who heard about the Bristol Pound from their friends in Australia! There has also been extraordinary interest from China.

Bristol Pounds went on sale at the launch. Serious numbers were taken as souvenirs, but, more importantly, many Bristol people took them for use. Members were still joining at the time, and are still joining today.

We have seen a gradual increase in text volumes from the start and, as of Jan 2016, we have sold a total of more than one-and-a-quarter-million Bristol Pounds since the launch. This is a good start, but it is only a start given the size of Bristol and its economy.

At this stage our sober estimate is that we have proved the system, but that we still need to establish it. We need to get the number of transactions to a significant volume, measured in millions of Bristol Pounds, and we need to expand our geographical coverage to take in all the very varied districts of the city.

Significant recent steps towards volume include: an agreement with our largest transport provider that the paper currency should be acceptable on the buses and at our local train station; an agreement with the local authority to accept domestic taxes in Bristol Pounds; the council beginning to procure services in Bristol Pounds; a partnership with Good Energy to accept Bristol Pounds in payment of gas and electricity bills.

In terms of geographical area, we have noticed that our first joiners were focused on particular areas of the city, where we have strong high streets with many independent retailers. Other areas have few shops and very little obvious presence of the local currency. They include some of the most disadvantaged communities in the city. In response we are just launching an initiative based on forming buying groups in these areas, which will source food and other products from local producers using the local currency. It is to be called the *Real Economy* and will be described under 'food systems' below.

Our links with other local currencies and their infrastructure support in the UK and Europe has been strong and positive. Our electronic payment system is based on Cyclos, a software package created by our European partners STRO and delivered to us by Qoin in the Netherlands, in partnership with the UK-based Brixton Pound.

Our European connections are continuing with a major new EU project in partnership with Sardex in Sardinia and a new local currency in Catalonia. The aim is to explore different models of local currency, including the provision of credit circles, to assess their potential impact on a local economy as they grow to scale. At the Bristol end this offers a new approach to volume trading and entails a new level of partnership working with the local authority.

At the same time we are developing the digital currency system so that it can be made available to towns and cities across the UK through a new 'Guild of Independent Currencies'. This will make it easy for any local town, or even a rural area, to set up a local currency to support their local businesses.

This project has been a very significant learning experience for me at many levels. As I point out, we are still some way from any sort of success in terms of long-term sustainability, and I am aware each day of our weaknesses, but there are some lessons that can be drawn even at this stage.

Firstly I sense that this project may be illustrating a new approach to the world of economics. It is a rebuttal of the *formulaic* economics that is dominating our society, and a restatement of a truer *narrative* economics in which economic activities are actively engaged with the beliefs, values and commitments of our lives and embedded in a narrative of progress that embraces the natural world. People work with the

Bristol Pound as an expression of their commitment to the local economy and their intention of working towards a sustainable city. Economic activity must be embedded in a rich narrative of human purposes. Only then can we make it meaningful and let it serve our vision of well-being.

Secondly, I have been overwhelmed by the sheer energy of the team of people involved in the Bristol Pound and have wondered how the team gathered and how that energy has been maintained. I will reflect further on this in Chapter Eleven.

Interventions in the Food System

I have been involved in a set of food-related enterprises in Bristol, which I will describe as examples of the different type of interventions that can be made in the overall quest to work for a sustainable city region.

There is a former vicarage, actually an old country house, now enveloped in a tower-block estate within two miles of Bristol center. Its garden is set in about a third of an acre and is enclosed by an ancient stone wall. Back in 2007/8, a few friends and I approached the church with a view to licensing it for use as a community project with the aim of helping the diverse people of this estate find a deeper relationship with the natural world through growing and cooking food. So it was that we launched what became known as 'The Walled Garden' project, serving the wider community with courses for school children and mental-health groups, with growing courses that teach English as a foreign language, a green wood-working initiative and various festivals through the year. Food is the centerpiece of this project, but it is not so much about the quantity of food produced, which is small, it is rather more about the process of interacting with food production and how this can increase our well-being as city dwellers. Our aim has been to develop a green oasis in the city where harmony with the natural world can become a felt reality.

At times we have spoken of our mission in terms of 'Soil, Soul and Society', which may be the best summary of our intent. The Walled Garden is just one of hundreds of such 'food-awareness' projects up and down the UK, encouraged so clearly by movements such as 'Incredible Edible' and funding provided by the Local Food fund.

Such 'food-awareness' projects should be distinguished, I believe, from enterprises that are serious about food supply. Around 2009, I became involved with a small group of people interested in community-supported agriculture. The council offered us five acres of land on a concessionary basis and we formed a project team to develop our plans. We had two growers, an accountant, a builder and others on our team, all committed to making this thing happen. Our interest was in a true community project, where our members did not just buy a veg box but shared the risk in the whole project. We worked hard at our values and our design, drawing as we could on Permaculture principles (more detail on Permaculture in Chapter Twelve). Sims Hill Shared Harvest, as we are known, has now run for five seasons and is feeding ninety households. Its members are consulted on all aspects of its life and it carries no debt. Having said that, it has also become apparent that food prices are really low in our society, driven down as they have been by the supermarket system, and that even the most well-motivated people are scarcely willing to bear the true costs of production.

In order to create more opportunities for initiatives like Sims Hill Shared Harvest to arise, it is necessary to pay attention to marketing and distribution of local produce. My sense was that we could create a much more direct interface between local producers and people in neighborhoods by forming buying groups that source directly from local suppliers via a web interface. Our inspiration was the *Gruppo di Aquisto Solidale* groups in Italy, which have proved that such direct relationships with suppliers can actively interest people in food and give growers a committed and substantial market.

There was also another challenge that we wanted to embrace. There are whole areas of Bristol where people have really no alternative but to get in their car and go to a supermarket. Local produce is scarcely available. These places are typically areas of high unemployment and tight budgets. They are also areas of Bristol where few people are aware of our local currency, the Bristol Pound. Our sense was that if we could align food supply with the local currency in a new initiative that would inspire interest in local food, encourage community development, personal budgeting, cooking from scratch and new enterprise, then we might be onto something. This project called *Real Economy Co-operative* is currently in its launch phase. Its aim is to set up 'Real Food Clubs' across the city, getting fresh food at affordable prices to people in the toughest circumstances.

These three case studies in the food system represent different types of intervention, each with their own unique set of outcomes and each of which may play a different part in working for a sustainable city. Other cities will have similar projects. They are all united in that they are not only functional in a simple sense in providing food, but each acts as a change driver in that it draws people together to achieve something while changing *perspectives* about food and deepening our *relationship* with nature at the same time.

Why Is This Chapter Important?

This chapter has proposed a broad strategy for dealing with climate change based on city regions and a process of localization. Its overall message is that there really is an alternative to the dominant systems that are currently so disruptive of nature, if we attend to the four key change drivers of finance, food, energy and governance. Each of these interventions has effects deeper than its apparent, instrumental goals, changing perspectives and drawing communities together around a new vision for human well-being.

So, it is possible to conceive of city regions creating a new type of money, whereby money is attached to values associated with sustainability. It is possible to imagine alternative energy supply systems, rooted in communities and benefiting from transmission efficiencies. And it is also possible to imagine a range of interventions in the food system to encourage local growing, awareness of food, of seasonality and of the soil, as well as dealing with justice issues concerning accessibility to local, fresh food.

The development of city regions as hubs of good practice in sustainability is also dependent on them attaining competence to act in development of their own life. This requires a shift in governance responsibilities from national parliaments to city regions.

Chapter Ten

The Fire Within

In a Nutshell

The great challenge of our day requires that we discover a deep connection with the natural world that moves us to give of our time and energy in this cause. This has both an outer and an inner component. The new approach to the sacred, proposed in Chapter Eight, finds it source in a contemplative relationship with the natural world that arises from the experience of nature itself, but is also nourished by paying attention to our inner lives.

We need to recognize the complexity of our inner lives. Very often there are deep conflicts among the beliefs, values and other commitments that we attempt to express in our actions. Some case studies will illustrate this. Reflection on our beliefs, values and commitments can help us reach a place of resonance in our lives. This is a place of inner integrity where our deep inner commitments exist in harmony with one another. In this condition we work with extraordinary energy, joy and commitment as we are expressing everything that we care about.

The Experience of Nature

There is no logical chain of thought or words adequate to describe the phenomena I want to bring before you here. We need to enter the realm of experience imaginatively and recognize that poetic descriptions may be more valid and useful than the analytic.

So I am going to use some words here to appeal to the reader's own experience of life. I am interested in what these words *evoke* in you, where you can feel a living connection and reality.

The first word is **unity**. I am interested in moments when you felt yourself a part of something bigger than yourself and bigger than a human collective. There is a certain sort of transcendence experienced by a football crowd, when their team scores, or by a choir as it hears itself, but I think many of us may also be aware of a greater unity, one that links us with the wider creation. You might have been standing on a headland, the wind in your ears, the waves rhythmically crashing on the rocks, the birds soaring in the air and just for a moment you feel part of it. It is as if you are no longer there, but you are joined into the whole scene before you. All the boundaries in your mind, all the checks and balances of your 'normal' existence, dissolve as you enter into this moment of participation in all that is.

The second word is **presence**. You are deeply present in this moment. There are no distractions, no phones, no worries. You are not so much concentrating, as letting go, relaxing into this place with all of your being. It is as if you are taking root. Drawing your energy from what is around, feeling its reality.

The third word is **creature-consciousness**. You may know yourself as embodied, feel your breathing, sense the reality of hunger, pain or physical pleasure, but you do so with an awareness of yourself as creature, joined in with all the other creatures of the universe.

The fourth word is **other**. There is a dynamic where we might be overcome by the sheer awesome power of the experience; feel humbled, even small, in the context of something or someone we cannot comprehend.

I believe that it is experiences such as this which are the true origin of our concern for the natural world and they are attested by all sorts of people, many of whom do not belong to a religious organization. For example, the TV personality and wit Clive James published a poem called 'Japanese Maple' in September 2014, just months before his anticipated death from leukemia.

The poem reflects on a tree that his daughter has given him and the hope that he will see it in its full autumn colors before he dies. What is apparent in the poem is that this man, nearing death, is appreciating nature with an unusual intensity and is comforted by the fact that it will go on when he is no more.

The poem speaks of his fading breath and discomfort, but he says:

...thought and sight remain
Enhanced, in fact. When did you ever see
So much sweet beauty as when fine rain falls
On that small tree...

It ends reflecting on the anticipated autumn color of the tree:

A final flood of colors will live on
As my mind dies,
Burned by my vision of a world that shone
So brightly at the last, and then was gone.

This poem illustrates an experience of life that is common to many of us at times. We can sense unity, presence and creature-consciousness in these words. There is no clear 'other' or appeal to a formed religion, but nevertheless there is a sensibility here that makes deep connections.

The poetic imagination is also clearly seen in the works of John Clare, from which I have quoted previously. Clare grew up in a village called Helpstone, near Peterborough in the early nineteenth century. People from London did not know what to do with this poetic genius, who was not schooled in their ways. They called him 'The Peasant Poet', and the great and good associated with him would tell of how much they had helped him to prominence. Yet John Clare was a genius in his own right and his writing illustrates a unity with, and compassion for, the natural world that few others have attained. He had spent his early life outdoors, and clearly loved every bit of it, taking paper and pencil with him and scribbling poems while pressing on his hat. His English was poor according to the rules of his day and he used many local words, but his vision and knowledge were immense.[2] Here is a short piece called 'The Beans in Blossom' that shows his sense of the complexity, unity and diversity of the natural world:

The southwest wind, how pleasant in the face
It breathes while, sauntering in a musing pace,
I roam these new-ploughed fields and by the side
Of this old wood where happy birds abide
And the rich blackbird through his golden bill
Utters wild music when the rest are still.
Now luscious come the scent of blossomed beans
That o'er the path in rich disorder leans.
'Mid which the bees in busy songs and toils
Load home luxuriantly their yellow spoils.
The herd cows toss the mole hills in their play
And often stand the stranger's steps at bay.
'Mid clover blossoms red and tawny white,
Strong scented with the Summer's warm delight.[3]

John Clare saw himself as situated in this wonderful creation. It was his home. It was his love and the source of his imagination. Some of his most famous poems fulminated against the process of enclosure through which he lived and the 'accursed wealth' that motivated the clearances.

Such relationships with the natural world are good examples of the new approach to the sacred introduced in Chapter Eight. It is clear that our relationship to the natural world has the potential, at least, to evoke a sense of the 'sacred', by which I mean not any formed religion, but a religious sensibility, a sense of that which is beyond us, or deep within us, and this is evidenced through unity, presence, creature-consciousness and the sense of other. This is not a case that can be made by argument. Readers must judge it for themselves, against their own experience. I simply believe it is real.

Another way of saying this is to imagine life itself as the breath, or 'spirit'[4], of the divine. Psalm 104 poetically describes the life of all creatures in terms of "thou sendest forth thy spirit, they are created" and "when they takest away their breath, they die."[5] This implies that the life we humans experience is a shared life, shared with all the plants and creatures of the Earth, and it is a gift.

So we have this sensibility towards the natural world, which, left to itself, without other influences, might cause us to ascribe a deep value to the natural world. We know we are part of it. We share its life. We sense its glory. Surely then we should guard it against abusers? Surely we should watch ourselves to check out our affect upon it? Surely we should long to live in harmony with it? And so there is a necessary moral dimension to the experience of holiness, but what direction should this take us in? The next chapter will explore some potential new expressions of this approach to the sacred.

The Fire Within

Such feelings for the natural world take their place within the whole world of our inner commitments. This is a place of enormous energy. It is the place where those things that really matter to us are held, defended and expressed in our lives. You can sense this in practice sometimes when you are talking with someone and the conversation suddenly goes to a new level. You might sense an energy or intensity arising. You listen with great concentration. You start to frame your words very carefully. There may be anger around, or great zeal. You realize that you are talking to this person about something that really matters to them. You have encountered their place of inner commitment. If you consider these moments in retrospect you will find that the moment that the conversation changed character was the moment that you began to tread on one another's beliefs, values or commitments. Sometimes the amount of energy you encounter in such conversations is too difficult to handle. A row ensues or an 'atmosphere' develops. Other times you may go away inspired or challenged.

This space is the very center of our being and source of meaning. We approach it with care as we might a 'holy place', such as a temple or church, yet it is within us. It is like a volcano, full of energy that can spill out destructively or a fire within that can be harnessed for good.

There are three dimensions to this space: our beliefs, our values and our commitments.

1. *Beliefs* are our theories about the way the world is and our place in it. We all hold some such beliefs. A scientist may embrace scientific theory as the route to a certain sort of knowledge about the material world. A religious person, or a person with strong political beliefs, may hold a different type of knowledge about the way the world is. Earlier chapters of this book have shown that these two types of theory, the one based on repeatable experiment and the other on narrative, need not necessarily conflict. Each person carries with them some theories about the way the world is. Some may not pay much attention to them and just 'get on with life', while others may spend their lives struggling with them and refining them.

2. *Values* are the things we care about beyond ourselves. We all have values. They may be expressed in a simple concern for our family. They may be about compassion for the poorest peoples, or a concern for those who are ill. They may be focused on justice or anger about oppression of marginalized

people, or love for the nation and loyalty between people. We have a whole spectrum of values that can extend to cover every aspect of life. To some degree they derive from our basic motivations as social animals (see chapter Two), but they are influenced by our moral development and the beliefs we come to embrace.

3. *Commitments* make up the practical, deep and organized feelings that we use to respond to the world on a day-to-day basis. In order to walk through life, in all its complexity and danger, we need rules of thumb, which allow us to respond immediately and effectively to the situations that we encounter. We simply cannot calculate a 'rational' response to every situation. Instead we develop an armory of commitments that determine our responses (see Chapter Four). These commitments can be influenced by our beliefs and values but often extend beyond them.

One of the most important things in any human life is to pay attention to this sacred space within us. It is the key to a fruitful life. We need to reflect continually on our beliefs, values and commitments and how they are working out in our lives. At one level this is a search for truth, not the dispassionate, abstracted 'truth' of the Enlightenment, but the full-orbed truth that measures feelings with fact and looks for meaning. At another level it is a search for integrity, whereby our beliefs, values and commitments are in tune with one another. In this condition we experience a resonance within and abounding energy, rather like the text of Isaiah which promised: "You shall rise up with wings like eagles. You shall run and not be weary. You shall walk and not faint."[6] Yet, sadly, for most of us, our 'sacred' space remains unexamined for most of our lives and often deeply conflicted within itself.

In this section I want to sketch some of the dysfunctions of our inner sacred space that can occur and the crippling effects that result. Problems can often be seen through analyzing the interfaces between beliefs and values, values and commitments etc. Our commitments are revealed in the way that we actually act in the world and, for many of us, they are a smorgasbord of cobbled together positions that may bear little relation to each other or the beliefs and values that we purport to hold.

I am going to illustrate this with three imaginary examples. They are not actual people that I know nor are they an attempt to define types of people. They are simply designed to show the confusion that can exist in this sacred space within us.

Example One:

A man attends a church in a prosperous city neighborhood. Since his youth he has been a keen believer, reads his Bible daily and belongs to a 'home group' that meets once a week. Much of this religious devotion is focused on the social environment of the church and on the theory of the faith. Very little pertains to his wider working life, where he feels a sense of isolation and little choice but just fits in to the culture around him. He works for a local firm of solicitors and generally keeps his head down and does not ask too many questions. Over the years he gradually accepts the basic commitments of that organization as they pertain to his work life and allows his religious beliefs to interact more with his family and home. He has little to do with his neighborhood except in church-led social ventures, many of which are actually thinly disguised 'mission' opportunities. He has recently taken interest in a food bank that has been consciously set up to express compassion. This is linked, in his mind, to belief through the stories of the Good Samaritan and the parable of the Sheep and the Goats.[7] He votes Conservative like his parents, but does not speak much about politics. He assumes that neo-liberal economics is the only game in town and supports moves that liberate the market from social influence. He is worried about the number of foreigners in the country today and views the past from the perspective of the privileged. He is quietly suspicious about climate change and does not think it is the business of the Church, where mission is the only true priority. In the end this world does not matter, it is getting right with God and our future in heaven that counts.

Comment on Example One: This man has an array of commitments including compassion for the poor (from a privileged perspective), political conservatism, distrust of foreigners, commitment to the free market and suspicion about climate change. Very few of these commitments have been examined at all. The religious faith that he espouses has an impact on a very limited section of his life and encourages him to ignore or bury other concerns that might arise.

Example Two:

This person was an ardent feminist as a young person and remains deeply committed to that cause. She is employed by the City Council Equalities team and has spent her life giving talks and promoting processes that ensure a full equalities agenda is pursued by the institution. Those around her would notice a very strong careerism in her make-up. She is always looking for a chance to shine among her peers and is acutely aware of power realities in her institution. Over the years she has adopted a language about equalities that implicitly makes others feel inadequate and advances her own career. Her home life is a struggle. She is now with her third partner and has two children from previous relationships. Resisting calls to marry, she tries to be very fair and compassionate to her children, but they do not see much of her and previous relationships complicate financial and social arrangements. She has been a strong Labour voter all her life and supports environmental campaigns, though as she grows older, her zeal in both these areas declines.

Comment on Example Two: This woman has an array of commitments ranging from a zeal for equalities, personal advancement, laissez-faire relationships, and a strong care for children. She has not examined her own desire for power or the manner in which this is pursued. She is leading a complex family life, dictated to by commitments about relationships and free choice. She is committed to environmental and social transformation and views these from a considered anti-establishment perspective.

Example Three:

A young person has been brought up by parents seeking a better world. She does not feel that she fits into society as it is and has spent her early years looking for an alternative way to live. She joins an informal grassroots environmental community in a city, embraces Permaculture as a belief system and looks to work it out in society, by way of building cooperative, inclusive and meaningful community; and in environmental terms, through getting stuck into land-based initiatives, growing food and building homes in ways that are in tune with the Earth. She also believes in Biodynamics and is ready to embrace the metaphysical theories of Rudolph Steiner. She works with relatively little money and consciously tries to make money of little importance. Projects that she engages with are of varied quality and endurance. Some projects fail through dominant personalities and much time is spent in learning techniques of conflict resolution. She is willing to work the benefit system to attain her ends. Little thought is given to her own future. She hopes her future will arise naturally from leading a meaningful life rather than planning deliberately for it.

Comment on Example Three: This young person has a strong set of beliefs that is being translated into values and commitments in a relatively coherent way. It may be that the beliefs tend toward romanticism and are being inadequately checked by science or a recognition of her dependence on the society as it is. There is the potential for these beliefs to give rise to projects that suffer from impractical idealism. This person may suffer from a series of practical disasters before her commitments develop a practical edge and become truly fruitful.

Exploring the Sacred Space Within

Only one of these three examples concerns a person with a mainline religious belief, yet each of these people has a set of beliefs, values and commitments that are very important to them. In other words, each has a clear sacred space in their lives. It is also true that each of them can obviously benefit from reflection on this inner space in their lives. There are some clear systemic failures evident even in these very short cameos.

The Christian in the prosperous church has a belief system that fails to engage adequately with whole sections of his life. In consequence he smuggles into his set of commitments a whole host of attitudes to life that cannot be criticized by his faith. They exist separately from it. This is an endemic problem in Christianity that reaches right back down to its roots. It concerns the influence of Greek thinking on the Hebrew mind,[8] whereby the very practical faith of Jesus becomes a theoretical construction encapsulated in a creed that barely mentions the actual life and purpose of Jesus of Nazareth. It also concerns the separation of the early Church from its heritage in the practical and Earth-related Jewish faith.

The Judeo-Christian tradition is particularly interesting for the way that it uses narrative to tell the 'story of God'. Narratives work by imagining the thoughts and feelings of others, so if God is in the narrative, then God is considered as a person with thoughts and feelings. It also means that thinking about God, conceived as perfect person, easily merges into thinking about what is true and right in itself and this could theoretically allow the faith to embed itself deeply into the society shaping its institutions and values.[9] Yet, in the example above, that was not happening. There is a tendency in some circles for talk about God to be too easy, for prayer to be 'familiar' to the point of absurdity, and for a general loss of a sense of the holy or sacred, which is at the heart of all religion. Another way of saying this is that Christianity, in its societal aspect, can become very close to a secular pursuit with little true religious element at all.

We have discussed in Chapter Eight how attributing humanity to God tends to form a human-centered religion. If people think of God in human terms, then inter-human issues come to dominate religion and its aims. As a result, it becomes all too easy for the religious person to conceive the Earth and its systems just as a stage for the great human drama and of little real value. The outcome of such a belief system is that environmental issues have no traction. Our faith adherent in Example One sees the primary concern of his faith as bringing others to believe. It is their relationship with God and their ultimate destiny that he cares about. Everything else is peripheral. It is for this reason above all that the Christian faith is failing to engage with the enormous and complex environmental issues with which we are faced today.

The most truly religious person in this set may actually have been Example Three. In my home city of Bristol, we have a burgeoning network of environmentally-minded people committed to discovering a new way of living. Their lives are often quite sacrificial in terms of what they are willing to go through for their ideals and there is a strong and very real sense of common identity. The question about such a movement must always be whether it can translate itself into successful practical action that will actually shift the society in the ways that are desired.

The most obviously secular example in this list was the equalities activist and in this case the person might well have objected to being placed alongside those with more explicitly spiritual types of belief. Nevertheless I would argue that, in the terms defined here, the equalities activist is most certainly demonstrating in her life a purposeful energy initiated by the sacred space within her. Her system of beliefs about the world are certainly shaping at least some of her values and commitments and thereby blazing a trail in the wider society.

The examples given in this section are, in one sense, extreme. Each of these fictional people has something they live for and this helps to make clear their sacred center. I am arguing that all of us have such a center. All of us have the potential to live lives of deep meaning because they are coherently connected to the fire within us. And each of us needs to reflect on our beliefs, values and commitments if we are to work towards integrity, wholeness and the true resonance that can occur when our sacred center is in harmony with itself.

Why Is This Chapter Important?

This chapter points to evidence that each and every human being has the capacity to recognize the sacred, both around them in their relationship with the natural world, and within them in the interplay between the beliefs, values and commitments with which they approach the world.

Many people experience the unity, presence, creature-consciousness and otherness that mark out a spiritual experience of the natural world.

It is also vital that we pay attention to the sacred center of our lives. Many people experience an unexamined clash of commitments at the heart of their lives. This drains them of energy and makes them ineffective people without integrity. If we can learn to find harmony within ourselves, we will experience a substantial resonance, a 'rising up', where our passions are truly aligned with our actions.

Eastern religions have a saying about God. They consider our inner life through the word 'Atman' – the soul. They consider the divine as 'Brahman'. Their saying goes 'Brahman is Atman'. This is a deep truth to which I am not sure I can do justice, but I think it claims that in some way attending to our inner life can be a route to the divine.

Chapter Eleven

Working Together with Purpose

In a Nutshell

Human beings are currently at rock bottom when it comes to living and working together. The dominant culture teaches us to be individualists and to be suspicious of community for the 'constraints' it might put upon us. Yet the great challenges of our day demand that we work together more effectively than ever right across the world. We must therefore rediscover the art of community life.

This chapter explores two avenues for recreating community. The first is the recognition of a set of five hallmarks that underlie deep social change. The chapter gives examples of how this might work out in relation to the natural world and building on the new approach to the sacred outlined in Chapter Eight. This includes a community song, an artistic installation associated with a new ritual of commitment to the Earth and a movement gathering people around a deep association with particular pieces of land as sacred.

The second avenue is based on forming new enterprises that unite people around values in action. This considers the idea of 'Shared Values, Clear Purpose' as the ground for creating the highest possible trust and co-operation in any enterprise. It also draws a distinction between doing a 'job', whose purposes we only half believe in, and doing 'meaningful work' that is in line with the commitments of our inner life. It suggests that making a transition to meaningful work can offer a whole new sense of well-being.

United in Something Truly Worthwhile

There is something mysterious about human community. It sort of happens, without being planned. It arises, we might say, from our very nature as social animals. Yet it can also be subverted, as is happening in our current society, by perverse individualistic perspectives on the world. Some years ago I wrote a research degree entitled 'Individualism and the breakdown of community life'.[1] It was about what the Enlightenment had done to our minds. The title still seems to strike a chord with people, perhaps awakening dormant longing within us for a stronger community life. I am not surprised by that, because I have sensed this longing myself and spent much of my life in search of a more meaningful community. It has not been an easy quest. There are so many false roads to tread. The small groups that become oppressive; the powerful leaders that dominate; the difficult people you cannot get away from; these are some of the perennial issues of community life. Yet in the last few years I think I have begun to experience something that has been an altogether richer experience that I would like to share with you in this chapter.

The starting place for me was a deep realization that we are living at a momentous time in human history in terms of our relationship with the Earth. And a sense that, if I am to live with integrity, I can do no other than devote my life to responding to this moment by searching for a better harmony with the natural world.

The exciting thing about this revelation was to realize that I was not alone. I became aware that there are thousands of people in my own city who share these concerns. Some are already living radical lives and looking for a fresh way of approaching being human beings on the Earth. There are the 'Transitioners', who were inspired by the Transition movement and joined into thousands and thousands of people across the world who are looking for initiatives and strategies around energy descent.

There are the 'Permaculturalists', who begin with the soil and attempt to translate natural abundance into a key to true human fruitfulness. And there are so many others who are not yet part of any movement but know in their bones that a new approach is needed and are looking. Over the last few years many of these people have become my friends and allies. Some are associated with our local currency initiative, where we have a community of more than two-thousand people, who see the potential to do something new in the economic sphere.

So that is my community. And what is special about it is that it is not organized, delineated, defined, in a conventional way, but it finds its coherence in a deep sense of purpose that binds everything together and creates a willingness to work together. Some people have looked back to communities in wartime and identified this same sort of spirit of working together. I have a sense that every human being on the planet needs to share this sense of purpose. Every human being on the planet needs to awake from slumber. And rise up. But what will make us rise?

The Five Hallmarks of Social Transformation

I believe there are five hallmarks of any movement that accomplishes deep social transformation:

1. *Imagination* – the first requirement of any social transformation is the ability to imagine an alternative to the present situation. We have lived through an age when it seemed like there was no alternative to the neoliberal economic consensus. We have been told that 'market realities' must dominate human societies, even that we have lived through 'the end of history',[2] because no alternative way of ordering human society exists. The truth is that all social change has been made by those who have believed and proclaimed that "There is an alternative." For example, the Civil Rights

Movement confronted racism in America and one of the most powerful moments in that process was when Martin Luther King stood up and proclaimed, "I have a dream." That sent a shudder of hope down the spines of all those who heard it. It was a bold assertion that things can be different. Anti-slavery campaigners took a similar tack; revealing the dreadful truths of the middle passage, they proclaimed, "There has to be another way." Women suffragettes likewise saw and proclaimed a world of gender equality, while today's environmental campaigners attempt to envisage a way of living in harmony with the Earth. Without our dreams we are stuck in our mess.

2. *Empathy* – true human progress may be best measured by the enhancement of our empathic abilities. One clear example of this in 2015/16 is the terrible refugee crisis that has swept across Europe. The sight of so many people, including families with children, risking everything as they set out from their destroyed homeland, moved us inwardly. It has helped us realize what it must mean to lose your home and your city, to be caught in the chaos of civil war, and to try to protect your loved ones. It may be that this refugee crisis is the true sign of globalization. And this is not a globalization of international companies or financial systems, but a sign of the enhancement of our empathic reach, so that we are truly aware of our impact on other societies and feel for the harm done. Such empathic sensitivity is a hallmark of every deep, progressive social change.

3. *Dedication* – an appeal for change must be connected to a moment of dedication or it can just fizzle away. I think we may be less good at this bit today. We need to create opportunities for people to say "I am in," "I will live this new reality, come what may," "I will dedicate my life to this movement." Such a moment settles something inside a person and joins them into a community of people who share that dedication. I remember

the time when I began to realize that the great challenge of the twenty-first century would come in terms of our relationship with the Earth. I could see the enormity of the change required in human behavior and the importance of it all. It was then I decided that *this* was something worthy of my life. And I felt enormous energy rise within me.

4. *Community* – communities form naturally among people who share a dedication to social change. Those who were on the march against the war in Iraq remember that they were there. The march was an act of dedication and, even now, people will speak about being on it. This is the stuff of community. People who share a concern, feel a bond. People who act on that concern are joined together by the shared act. The articulation of a clearer sense of community may be crucial today.

5. *Celebration* – one way that communities find their identity is by means of common celebration. My city of Bristol is known for its festivals. Hundreds of thousands of people turn out each year to enjoy the Harborside Festival, the Balloon Fiesta and others like it. And in celebrating they *feel* part of their city. Our own Bristol Pound holds a night market each year on the anniversary of our launch. There is music, street theatre, food and general mayhem. Loads of people come and the celebration does something to our sense of citizenship. People feel proud to be part of such a vibrant scene. They sense something good about a city that works together creatively. They grow in hope for a better future.

These five hallmarks of social transformation are all found in effective movements to some degree. You can see them in the political rally with the big speakers telling a story that inspires imagination and empathy. You can feel the dedication as the gathered company join in a traditional community song. It releases a sense of community in amongst the celebration.

These hallmarks are also formally present in many religious gatherings, where a speaker might attempt to evoke imagination and empathy towards change, where the community may have established initiation rituals, common meals and regular celebrations. My sense is that those of us who long to initiate social change might do well to reflect on these hallmarks of effective social change.

Here are a few ideas that have come to me about things that could be done at this moment to express the movement to live in a better harmony with the Earth.

A Song for the Earth

A few years ago Coca Cola did an advert based on a song which opened with "I'd like to teach the world to sing in perfect harmony…" I remember that this advert made me feel good. Of course, that is what the advertisers want. Yet in this case, I think I caught a glimpse of something bigger. What if the world really did sing together? What if the world did sing about something that really mattered? Could it inspire us? Could it draw us together?

This summer my wife and I treated ourselves to a cruise on a tall ship. It was our fortieth wedding anniversary and time to have some fun. Each evening as the ship set sail from a port in the Mediterranean, we would all gather on deck. The anchor would be lifted and the great sails unfurled. As the ship felt the breeze so it would begin to move through the water and music would be played to us gathered on deck. It was music I had never heard before, a Vangelis composition, but it fitted the mood, and we were caught up in a great sense of awe, sailing this tall ship with five masts and forty-two sails, drawing through the waters with the land disappearing over the horizon.

Such experiences of the power of music in a natural context make me wonder about looking for 'A Song for the Earth'.

In Bristol we have several large community choirs, performing a wide repertoire of songs. And the choir members love it. Music is deep within all of us. Ian McGilchrist proposes that music comes prior to words in the story of our development. As such it is a conduit for an expression of shared feeling, with the potential to express collective emotions and to form a gathering point for communities (see Chapter Three).

Music seems to derive its community function from the act of mimicry. To tune into the music of others is at the same time to tune into their emotions and their deepest concerns. And we are lost in the *communality* of the process. Now let's translate this into the present moment and ask, "What would happen if the world were united in song and this song was a song for the Earth?" Can you feel the potential power of this expression of our current purpose and mission? A song expressing our longing for a better harmony with the Earth and thousands of people in every city of the globe united in singing it. How would that be? I think that could unite us in changing the world in a deeper way than all the conferences and leader statements put together.

I would love to see cities and peoples across the world coming up with their ideas of a song for the Earth, with each place creating opportunities for thousands of people to join together in song. How easily our media could then grow this phenomenon. This would not be centered on performance, but on participation. It would not be competitive but appreciative. And it would bind the world together in response to the natural world.

A Moment of Dedication

We may be missing something in our search for a new purposeful community life. It is something that has been known to religions across the world throughout human history. This is the moment of commitment based on a ritual action.[3]

Many cities have central spaces, where people naturally

gather for festivals and rallies. They are places that are felt to be the heart of the city. They connect with history, with how the city feels about itself. People identify with this space. They walk through it. They gather there for a festival or a rally. It is their space and symbolizes their city.

Now how would it be if the city's artists conceived something for that space that was inspiring, even shocking, in the form of an arch, which people were invited to walk through as a symbol of personal dedication to living in harmony with the Earth?

I really don't know what it would look like. It doesn't even need to be permanent, but it would be inspiring and stand as an invitation to commit to the Earth. Some have said to me that it should take the form of a tunnel to emphasize the process of going through something. Others have suggested a bell is rung by each person who goes through, or perhaps a photo is taken and put on social media, so that we create a wave of commitment across the Earth.

Any city serious about its relationship with the Earth could do this. It could be supported by events around the arch, such as the song above. It could become the beating heart of a city's concern to become sustainable, where action was not just down to leaders and policymakers, but where all the citizens came to embrace the process.

The Sacred Land Movement

This is a slightly different take on the idea of dedication, where people are invited to make a bond with a specific piece of land. We have unfortunately become accustomed to thinking about land in terms of private ownership rather than collective responsibility. This has the dual effect of alienating ordinary people from the land and diminishing our appreciation for the natural world.

The Sacred Land movement is an idea to counter this.[4] It

invites us to consider land close to us and look to designate appropriate places as sacred. Any piece of land can be designated as sacred if there is a commitment made by a set of guardians,

> To cherish all that lives within it;
> To maintain a rich and diverse balance of species;
> To remove all human debris;
> To exclude all use of poisons.

The guardians, who would meet at least once per year on the land, affirm these vows, maybe in the form of a celebration, and decide how they would fulfil their vows in the coming year.

There would be a minimum of ten guardians per designated area of sacred land, with the number of guardians increasing with the size of the piece of land being cared for, so as to always have at least ten guardians per acre of land. One person would only be allowed to be guardian for one designated space at a time and a public list of guardians would be kept.

Such active engagement with the land might be appreciated by many public bodies, which are looking to involve the public in helping maintain parks and landscapes. An agreement as to the demarcation of responsibilities would need to be drawn up with the legal landowners, but the promise of this movement would have the potential to interest many new people in actively caring for the natural world, coming to appreciate the intricate balance of ecosystems in which we participate and the effect of human beings upon these. In some subtle ways it would be renewing our collective responsibility for land itself, something that was gradually eroded through the enclosure movement. Prior to the enclosures in the UK, most ordinary people had land that they worked for themselves, while earning cash and paying taxes by means of a supplementary trade. Through this they were able to maintain a vital link with the land and its well-being. We often do not have this today, especially those of us who dwell in cities.

The Sacred Land movement could add something distinctive to a variety of existing initiatives. So, for example, a community-supported agriculture project might join the Sacred Land Movement and its own members become the first guardians of their piece of land. It could also use the Sacred Land movement to recruit new members to itself and encourage engagement with the area of land under cultivation. It could ensure biodiversity on the site by embracing Permaculture Principles and making this part of its presentation to the public.

Alternatively an existing 'Friends' group caring for a park could become part of the Sacred Land Movement. Those who regularly pick up litter might be honored for their fulfilment of the vows and the other members could take an interest in creating wilder spaces and enhancing biodiversity on the site. The members might set out to catalogue 'everything that lives' on and around their land, using a mix of social media and expert help. The project could involve young and older people, cultivate appreciation of diversity and work in tandem with scientific understanding.

The Sacred Land movement might also help some groups campaign for land to be kept or restored to its natural state. The land could be designated as sacred as part of creating a mass movement of guardians and others with a real interest and commitment to it. This would be so much more meaningful than petitions and other forms of traditional campaigning. The number of guardians would be an active measure of public engagement.

Meaningful Work

Now I am going to change tack a little, but with a similar intent to what has gone before. I want to begin by asking a question of us. Many people go out to 'work'. They have a 'job'. They earn money, pay the mortgage and otherwise exist, but do they really *live*? It seems to me one of the greatest privileges of life is to be taken up in work that you really believe in and one of the greatest tragedies is just turning the handle to provide for your subsistence. How then can we make the transition from a regular 'job' to 'meaningful work' and how does that feel?

I described above how I came to feel that the overwhelming issue of today's world was that human beings have to find a way of living in better harmony with the Earth and all that lives. When I recognized this, at the age of around 55, I knew that I had found something that I could devote my life to. I believe that getting in touch with our deepest motivations and values is the key to finding meaningful work, but this is not necessarily via the traditional 'career' methods with training schemes and formal job applications. An alternative is to just get stuck in.

It may be that you will form a new enterprise, or you may join an existing one. I found that I was able to go part-time in my main work, so as to devote time to new projects, like the Bristol Pound, without taking any money. This attitude proved to be common among the Bristol Pound team. Almost all our paid staff at Bristol Pound started out as volunteers and have gradually made a transition to paid work as money became available. It has made for one of the best motivated teams I have ever worked with. It seems that the secret of meaningful work lies in following our values and not being tied strictly to reward. Such attitudes foster community life, encouraging others to return favors and live in a more gift-like economy.

Shared Values, Clear Purpose

This type of thinking can also be applied to the enterprises that we are forming, or long to form, and can help our society make the transition it needs to make at this time. Values are where our energy comes from. They are from the fire in each one of us. So it is that being clear about *shared* values may be the most useful way of discerning who we can work with on a project. The Bristol Pound formed very slowly. We were three years before it launched and spent a whole year simply doing a feasibility study. This meant that the small, emerging team spent a lot of time together. We had no office, so we held a succession of coffeeshop meetings and we talked about all sorts of things. When any group of people tells stories about themselves, they show what they care about. We subtly declare our values. For our team, climate change and the environmental issues we face as a society today were a central concern. We knew that society had to respond and we wanted to precipitate that change. We also cared about our local area. We wanted to promote the vibrant community life of our city, the inclusion of marginalized people and arrest the decline of our high streets. These were some of the values we shared. At the same time we were also very different people, bringing different skills and aptitudes to the mix and, as the team grew, this diversity increased.

The other thing that united us was the clear *purpose* around which we were co-operating. Lots of the detail was being developed, but the core agreed purpose was a local currency, at scale, across the city with paper and digital elements. We were inspired by the pioneers of local currency in other areas like Totnes, but sensed that, with a city, there was a chance to take this to scale.

To unpack this a little further, what was important, I think, was not that we shared values about everything associated with our lives, but that we shared values that related to the project on which we were focused. We were, and are, very different people. There are parts of our lives where we would most definitely not concur about values, but we are confident of one another in the areas that relate to the project we are pursuing together.

I have personally experienced the importance of Shared

Values and Clear Purpose in more than one project. The community-supported agriculture team that I was part of worked together for two years before its launch. We spent six months talking about what we were going to try to achieve, who would benefit and what values the project was based on. The process was really important. By the time we came to launch we had thought through almost every aspect of the scheme. People were impressed. Sensing our values and clear plan they were keen to join the community and willing to invest enough money to get us through the first growing season.

I have also experienced these principles in a negative sense, as in the plethora of committees in our society that have no clear idea of what they are trying to achieve and little shared value base. Having been involved with churches for years, for example, I know just how dysfunctional the church committee can be. *The Vicar of Dibley* was not far off! Churches have worked from a position that what they needed to attend to was simply shared *belief*. They have been less aware that shared belief does not always translate into shared values or common commitments out of which we live. There *are* shared values among Christian believers but they tend to be confined to particular areas like compassion and kindness, honesty and integrity. These are important components of the fabric of any society, but they inspire a very small subset of projects. It means that great fault lines regularly appear in important conversations on anything from reordering the inside of the church to action on debt because the church members simply do not care about the same things. They have also often failed to do the serious work of building a common understanding about what they are trying to achieve together. My sense is that a large percentage of committee life in this country is currently experienced as a waste of time and energy by the participants. This diminishes participation in society.

This brings me to a difficult practical issue. I have heard from around the country about many projects that fold up. People may have become burnt out. Internal conflicts may have prevented anything getting off the ground at all. Dominant leaders might have developed that split the group. The whole idea might have run out of steam. There are a host of potential pitfalls and, sadly, some of the most enthusiastic lives are scarred by these events. Here are some ideas about keeping projects on track and our lives in balance. They derive from my own experiences working with others to achieve social transformation.

1. Retain focus

The biggest danger in regard to burn out may be loss of focus. Don't try to do everything. Resist strenuously the person who wants the project to develop down a whole new pathway. Define what it is about and make that *all* that it is about. If someone does not want to do this, they will go away.

2. Work in a team

Projects that are initiated by just one person are terribly vulnerable. Try to create a stable project-development team where you are spending time together focused on the task. The task is the agreed place for expression of your energies. These conversations will also be an unstated and even subconscious means by which you assure each other of your shared values and so build trust. New people will join this as the project grows, but beware of too many people drifting in and out. Communities need stability in order to build effective trust.

3. Leadership

One of the great advantages of shared values/clear purpose is that it allows a group to work consensually and gets away from the standard dominant, defined leader models. Because you trust one another in terms of the basic project, you can genuinely listen to different ideas brought in by members of the group. That way you will be most creative. This will tend to favor co-operative forms of governance for the project as it develops.

4. Allow the project to mature

Projects go through a life of their own. In the Bristol Pound, hardly anyone got paid for the first three years. Then we were able to pay two staff, part-time and a very low wage, but it was something. Then, when some serious grants came in, we could all get paid. At the same time, through this process, we had to decide what we were going to be in terms of a corporate model. We had to begin to define roles, designate directors, set salary rates and generally begin to form up into an accountable body. There was no alternative to this process. We had to be responsible about the operation or the public would not have trusted it. That meant that we needed to set up transparent and accountable structures. Sometimes people at the more idealistic end of the spectrum try to avoid structures altogether. Unfortunately this leaves them open to losing public confidence and prey to dominant leaders, who simply take over. The maturing of a project that has depended completely on voluntary energy can be a real trial and is not easy to do well.

Maturity also requires financial realism. A project might begin with enormous voluntary energy, but it must become sustainable in every sense and that will normally mean paying some people. Some people, who are very clear in regard to values, are very weak in respect of financial realism. This is a serious difficulty for many projects. A realistic business case needs to be built into the project from the start and progress must be checked against this, so that when the grants run out, the project continues. This is easier said than done.

5. Become aware of your own skills

As your project matures you will become more aware of the particular contribution you are able to make within the team. A project in its initial stages tends to have everyone involved in everything. As it matures, so the different aspects of delivery are delegated. There are different tasks and roles and responsibilities and you will not be involved in it all. That is a good thing and an opportunity for you to be clear about what you can specifically offer.

6. Ground yourself

The earlier chapters of this book are designed to give a philosophical and emotional context for your work. They are an attempt to help you reflect deeply on the state of our humanity today, the strengths and the weaknesses of human society and the enormous social and environmental issues bearing down on us. More than that, they are designed to give you the tools to think about your own practical projects and how they begin to address these issues. It is all too easy to be overwhelmed by the state of the world today, but if we can connect our own projects with the big picture and understand what we are doing in that context, then we can say "I have seen the needs of the world. I have felt the needs of the world. I know I cannot do everything, but this is what I am doing and why." If such a statement can be a conscious part of our life, then you are much less likely to burn out.

7. Risk failure and know when to withdraw

Every transformative project is a risk. Some are very public risks in the sense that your reputation will be at stake. There will be many moments when the project teeters on failure. And the great danger is that you will then personally step in, take responsibility for it and try to cover its weakness. If you are genuinely working in a team, then the team should be held corporately responsible for the project, not you personally. The team needs to maintain a difficult balance between a realistic awareness of weakness and a resolute confidence in the possibility of success.

The emphasis on team-working should also mean that you are free to leave when you need to. This is very important. Burn out comes to those who feel trapped in their obligations. There is a moment when you have done what you can do for this project. It can go on without you. It may not feel like it can, but actually it will survive, and even thrive, if you withdraw. There is also a moment when some of the initiating team *should* withdraw for the sake of the project. The energy and enthusiasm around the launch of a project can be very different from the mature delivery phase and the people required for each phase are usually different. We need to recognize that.

Why Is This Chapter Important?

We have to learn to co-operate better if we are to address the great challenges that we are faced with. This chapter has explored the possibility that the key to renewed community life is a common sense of purpose. It has described five elements that lie at the heart of all deep social change and offered examples of how these might be worked out today, including a new approach to the sacred that expresses purpose and vision for a better harmony with the Earth. It has considered the phrase 'Shared Values and Clear Purpose' as the basis for forming a team to deliver a new enterprise. The evidence suggests that attending to these things can form enterprises of the greatest trust and co-operation with the minimum of hierarchy. This chapter spells out some ways in which people may find a transition to meaningful work that chimes in with their deepest values, by seeking out how they can contribute to social transformation.

Chapter Twelve

Finding the Path Ahead

In a Nutshell

We are people on a journey. We are joined on this journey by the millions of other human beings that have lived on this planet, but this generation bears a particular responsibility. This chapter draws some of the threads of this book together as a way of considering our own journey through life and what we can offer to the world. It commends a move from being passive recipients of culture to active makers of culture. It recommends serious engagement with creativity, storytelling and artistic expression. And it encourages the embrace of three disciplines, the discipline of imagination and reflection, the discipline of engagement with nature and the discipline of social transformation.

Coming Alive

There is a passage in the otherwise rather weird book of Ezekiel where the narrator is taken in his imagination to a valley in a hot, deserted place that is full of dry bones. And the word comes to him "Speak to the bones," "Tell these dry bones to live." So Ezekiel speaks to the bones, and he stands amazed as he sees in his mind that the bones begin to move, even to 'rattle' as they rise up, and flesh forms around them and they are alive. For him it is a picture of the renewal of the society of his day.

Ezekiel's vision sums up, for me, something of the wholesale change in our culture that we are looking for today. There is something strangely passive about the lives of people in Western society. Very few people are really trying to change anything. The great movements of the past seem to have faded away. It is like we have come to believe that nothing really can be changed. We must just watch telly, eat, drink and consult Facebook. Likewise politics seems to have become the art of keeping people happy. One of the greatest signs of this condition appears to be the widespread acceptance of the description of human beings as 'consumers'. How did we come to accept this title? What trick of our minds reduced us from 'active citizens' to 'consumers'? Surely it is time for us, like those dry bones, to rise up with life and become active agents of the change we know is possible?

Ian McGilchrist[3] says we are currently a myopic culture based on the dominance of a 'left-hemisphere' approach to the world. The left hemisphere pays attention to the world with a particular 'focused attention', defining language and the world in terms of a set of objects and categories, rendering great power to the society, providing a scientific description of everything, even that which it should not so describe, but failing to embed our perception in the holistic frame that is offered by right-hemisphere thinking. As such we fail to engage properly with our emotions and the whole set of cares and responsibilities that we need to embrace if we are to find true well-being as creatures on the Earth. This concluding chapter tries to open up ways of renewing our 'right-hemisphere' appreciation of the world, so as to counterbalance the dominant culture in which we find ourselves. In so doing it aims to help us tap into the energies that come from an alignment between the inner core of our lives and the outer world of community and society.

Setting Out on the Journey

I was brought up with a very scientific head on my shoulders. In my younger days it was the way I found easy. I could get right answers. All that fuzzy stuff like history was harder, altogether more complex and, so I thought, best avoided. Yet, as time went on, I was forced to face up to things from which my young life had been shielded. I met people caught up in chaos and tragedy from their earliest days. I met others in desperate poverty and need. And the easy answers fell away. I had to find a different way to think and to live.

I remember a time when my head was full of the importance of narrative to society. I had realized, as I describe in Chapter Three, that narrative truth is a key component of every good society. Then I met someone who said to me, "Chris, you talk about narrative, you write about narrative, why don't you start telling some stories!" And that set me going. I began, hesitatingly at first, to try to tell stories. I was particularly interested in stories that made people think and which would generate discussion. Yet, even as I began to do this, I realized that it was tapping into a completely different part of my mind. Telling stories was emotional. I was not used to being emotional. It was also engaging. You had to look people in the eye, respond to their response, and bring the whole thing alive. It was so different from the abstracted thought that had dominated my life until then. It was also an emotionally healing place. I recall how early on I used to tell lots of really dark tales, about people doing ugly things to each other. (They were also often biblical tales, but that's another story.) Yet the time came when one day I woke up and realized that I did not need to tell that sort of story any more. I could be positive, optimistic and hopeful. It was a great change.

I mention that process because it may be that others need to release the creative side of themselves in some way. Storytelling is a fascinating thing to get into, but others might be drawn to singing in a choir, playing music, or perhaps making art or sculpture. The great thing about all these creative pursuits is that they are irreducibly holistic. They require us to bring all of ourselves to bear on the pursuit, our emotions, our experience, our skill, even our vision for life itself. This is surely the renewal of those right-brained processes that McGilchrist commends.

Such things are important, not just for their own sake, but for what they do for our well-being and for the well-being of the wider society. Many times in this book I have implied that things cannot change because the great mass of people do not wish them to change. So, the politicians cannot lead because we do not want to follow. The bank leaders cannot reform the banks because the shareholders don't want it to happen. The journalists can't discuss the truth about climate change because most people don't want to hear it. So it is that there is a great need for transformation in the minds of the general public if the human being is to live well on this planet. The pathway to such transformation may begin with personal creativity and the healing that comes from that.

Walking the Walk

The Santiago de Compostella pilgrimage is an interesting phenomenon. Something over two-hundred-thousand people walk stages of this walk through France and Spain each year. And for many walkers, this is not so much a religious quest as a personal search for meaning and healing. Walking through nature is a rather wonderful thing. Your speed is strictly limited by the paces of your body. You are walking in the natural world, but it too is moving at its own pace. You become participants in that world, fellow travelers with the birds and in communion with the plants. There is no rushing here, no appointments to be kept, just the long path ahead and your companions. There is probably nothing that unites human beings down the ages as much as walking. We walk with everyone who has ever walked. We walk with the creatures that have always accompanied the people. We see the plants and trees that have been there for millennia. We are part of something very big. And it is all good for us.

I think such a journey is a really good metaphor for our lives. We walk from scene to scene in our lives. New places bring new challenges and new opportunities. As we go along we are always changing in a process that never stops until we stop. Every day we can be changed inside. Every day we can set out afresh. Every day we can commune with the amazing world around us. Such is the walk of life.

I would like to suggest now a few disciplines that will keep us on a fruitful path.

The Discipline of Imagination and Reflection

Attending to the sacred center of our lives is the first vital element of a fruitful life. The heart of this discipline is a regular time put aside to settle ourselves, become aware of our lives and the pressures on us, and notice how we are responding. Some would call it mindfulness. It leads to an examination of the beliefs, values and other commitments that we are working with. It implies a rigorous openness to the truth about what we hold dear, a willingness to accept challenges right into the heart of our being, to learn and be transformed. In a sense it is to move from a castle mentality about our sacred center to seeing our inner life as a developing ecology. The natural default position of many people is simply to defend our current set of beliefs, values and commitments. Yet people who make imagination and self-reflection their disciplines will cultivate a radical openness to all that is true about themselves, in relation to human society, in relation to the Earth and all its other creatures and the divine. Such discipline is not for the cowardly, but it is the path to transformation and change.

Some would call this discipline 'prayer', but it is prayer without a formula. It is a reaching-out from within that can take the form of everything from formal prayer to expression in the creative arts. I am persuaded that poetry, music, art and story-telling are important ways of learning to say something beyond conventional conversation. They are disciplines that come from deep within and these arts need practice if they are to bear their fruit and help us find our way.

Community life should also be seen partly as an act of self-reflection. Even the simplest communal act, like taking time with a friend over a gatepost, involves noticing how they respond to us, expressing ourselves in words, manners and gestures and noticing their response and, in so doing, we build bonds that are vital to community life. The recovery of quality face-to-face communication is a vital need of Western society. It is too easy to believe in the false representations that we make of ourselves on social media. It is too easy to live a superficial life on the internet. It is good simply to take time to be with people.

The Discipline of Being in Nature

I believe it may be good to formalize what has been happening all over the world, namely that people have discovered that being outside and being in the natural world is good for our well-being. Some may walk, others may sail. Some will climb mountains. And we all feel strangely better for our contact with the natural world.

In the last few decades thousands and thousands of people in the West have explored growing food for the first time. To touch the soil with our hands and observe the growth of plants is to mysteriously bind ourselves to a common future. To grow our veg may be the first time we have really cared about the well-being of a plant. As years go on we will come to rapidly sense the first signs of lack of nutrition as the plants' leaves lose their color. We will spot those first dreadful signs of the root fly at our cabbages and know when the birds like to take our fruit. It is a wonderful and richly rewarding journey that takes us beyond the veg to the well-being of the planet, beyond the raised bed, to the atmosphere above. Becoming aware of the impact of the weather, noticing the birds, the insects and the whole variety of life forms, even the slugs and the weeds – these things root us in the sacred.

The place in which we work with the Earth will become known to us in a quite different way from any office we may attend. It will be experienced as an open space, where careful observation makes for surprise and delight and new kinship connections with the Earth and its creatures will be forged. For some this place will become sacred to us, a place of contemplation and peace.

Tending the Earth also ensures our focus on food as the very energy of our existence. Being aware of the fragility of food systems is a vital lesson we need to learn. Building more resilient food systems is one of our greatest challenges. Tending the Earth is not simply taking recourse to Eden and a romantic ideal, it is paying attention to the need to work in the 'sweat of your brow', to be able to withstand great crop failures and the human pressures that will result. That is why tending the Earth is something so many are beginning to appreciate.

The Discipline of Transformative Action

The days stretch out before us. None of us knows how many we will have. But, at the end, every one of us will want to feel that we have done something worthwhile. There are those who drift, maybe earning a lot of money, maybe seeking pleasure here and there, but who never find real purpose. The watchword of such existences is boredom. People get bored seeking their own pleasure.

This book has pointed again and again to our need to find a better harmony with the natural world. This is not an option for humanity. It is vital to our future. As I understand it this is the great focus around which the purposes of our lives can find their fulfilment. There are so many things that we can do, both as individuals and joining together with others, to address the issues that we are faced with. My own journey has been a wonderful experience. Beginning with our first feelers about solar water-heating, then solar PV on our roof, clearing our lane and planting fruit trees, building a wooden conservatory, joining in a project to help transform gardens to growing places, helping to start a community-supported agriculture project, forming a local community orchard, helping pioneer the Bristol Pound and now a food cooperative... The list is long, but it has all been about transformative action. There are so many things that can be done. So many ways that projects can be approached. Yet the extraordinary thing is the energy that can be discovered to do these things, when they are really in tune with our deepest values. I remember sitting in my conservatory the other morning and thinking that I had come to live in a sort of paradise. Perhaps that is the reward of following our own inner urges towards transformative action.

Feeling our Relationship with the World

It is time to reassert the importance of feelings in our shared lives. For too long the rationalists have held sway with their call for us to subdue our emotions and work only from material evidence. I hope this book may have convinced you that a rich life is full of stories that communicate feeling; but how do we begin to establish this in our public life?

As described in Chapter Two, the origin of many of the feelings of social animals, like ourselves, can almost certainly be traced to the care of the young. This is where, in terms of evolutionary history, we learnt to understand the feelings of another person. Such primordial empathy is the root of much of what we truly value in the world. It could be that the true 'measure' of a society is its empathic reach. The perennial problems that echo down the ages, about inclusion of the marginalized and our approach to the one who is perceived as 'other', may actually hinge around our empathic ability. Can we see this person as someone like ourselves with a similar set of feelings? If so we will develop a system of morality that informs public policy and ensures that they are treated well.

The same question is even more pertinent to the natural world. We simply must come to *feel* for the well-being of nature. We must *grieve* when nature is despoiled. We must *love* its variety and complexity. We must *long for* our world to adapt its ways so as to live more in harmony with the Earth. If we cannot do these things, then we are truly lost and human beings will suffer catastrophically in the coming century.

Grief, Loss, Failure and Other Opportunities

Most of us who have lived any length of time have known personal grief. It comes in many forms, the loss of someone dear to us, the failure of a project we have poured our lives into, or even a chronic, creeping depression that dogs our steps for years. You will know the symptoms of this grief and have your own particular way of coming to terms with it. What I would like to recognize here is that grief is a rather special kind of energy in our lives that we can use to some degree.

Grief sets us up to see the big picture. It sets us apart from the mundane aspects of our existence, the day-to-day grind of life, and forces us into a bigger world of emotion and of nature with a renewed search for meaning. At times it seems that meaning has completely eluded us, everything is now hopeless and there is no point to anything, but if we can hang on, something new can happen.

When I was young we used to play with kaleidoscopes. They were cardboard tubes that you looked into and saw a pattern made by crystals. The game was to twist the tube and see the crystals form a new, and often surprising, pattern. Grief, as I see it, is rather like twisting the tube. It changes the way we see the world and allows a new pattern to form. I have made some really big transitions in my own life in terms of career and I don't regret one of them. They have all opened up new opportunities and become a form of enrichment. Yet I am aware that sometimes I needed shaking out of the thing that I was in. It would have been too easy just to stay and go on doing what I had been doing. I had to move on and it was grief, I believe, in one or two important instances, that forced the issue and made me look further, deeper and more radically than I would have otherwise.

Recognize Life beyond Words

Coupled with the impulse towards emotional sensibility is the recognition that there is a vast realm of life that we simply do not describe in words. As other social animals negotiate their complex lives without using words, so we can do something really quite similar. We still have abilities to communicate at that level, but we have forgotten their importance and they have become attenuated through lack of use. The touch, the glance, the smile. They do things to us. They are important. The community, the team, the family. They all have their own reality, which is sensed and known by us in a way that is not expressed in words, and can even be *lessened* by the use of words.

It could be that the route to a true spirituality begins with a sort of going back to our experience of life prior to language, when we felt things about our social lives and communicated in all sorts of rich ways, including body language, signs and rituals. Try keeping silent with someone for a little, or just using fewer words, while continuing to gently communicate. Aim to live the simple things, without words. I anticipate a deepening of your relationship with that person and a pathway opening up to a new level of experience.

Creating a Values-Laden Language

There is a need to change the *type of discourse* in society. We need to move from the formulaic, static and myopic version of purposes that currently keeps us in its cage, through to something far more imaginative, holistic, experiential and evolving. The way that we can achieve this is by centering our public discourse on values rather than policies. Policy-led language tends to be directed towards utilitarian ends. It is about "How many people will be better or worse off under this tax regime or that?" Or "How many immigrants will come in if we do this and how much of our benefits pot will they consume?" That sort of language will never get us to raise our eyes.

To talk the language of values is to engage with what really concerns us. If we speak and live out of our values we will find that we have enormous energy. All that great depression of meaninglessness, which hung about our lives and weighed us down, will be lifted. I have given some examples of how a values-led approach to economics might transform our perception of finance, or our food systems. Others could tell tales of innovative energy systems that have arisen from people who care about the planet's future and the transition to sustainable energy. These are all values-led people, who have found a practical way of expressing what they care about. The strongest example of a values-led movement of this type to date is the Transition movement. Transition is now global in extent with thousands of projects that have grown out of its embrace. And its attractiveness, I would contend, is that people have been drawn together around shared values related to the future of the Earth.

As mentioned previously, there was a time when people thought that the way to deal with global warming was to encourage personal behavior change. There were several problems with this. Firstly it did not adequately focus on the structural elements of the problem, like the global financial system. Secondly the government's advisors thought that the way that this could be achieved was to treat the whole project like a marketing exercise, segmenting the population into different types and interests and playing advertising to those interests. Some people were deeply uneasy about this method and that unease gave rise to 'Common Cause' (introduced in Chapter Four).[4] The killer conclusion of their analysis was that subjects like caring for the Earth actually *require* us to be 'other'-regarding and so will be undermined by the sort of appeals to self-interest that government agencies were suggesting at the time. To engage with the issues of global warming *must* involve us working with values beyond ourselves. I see the importance of Common Cause in helping to give us a language, acceptable to religionists and secularists alike, through which the global community might discuss what it truly cares about.

Nevertheless values are a tricky subject, because they are actually rooted in our feelings and are based on our empathic sensitivities. This means that they resist language to some degree. They are part of that reality of our experience and commitment that is beyond words. This is partly why human beings have developed art, poetry and storytelling. Music, as mentioned previously, could also be a really important part of the transmission of values. All these art forms can transmit feeling in its raw, undefined state and appeal to others on that basis. For this reason I suspect that values discourse will be far less defined than the pretended rationality of modern public life. I want to hear about our relationship with the Earth in ways that evoke our passions. I expect to be shocked, struck down in amazement and brought to my knees through the power of communication as the importance of our values in relation to the Earth is impressed upon us in a whole range of ways.

Rise Up

I cannot presume, but I hope, that some readers are ready to try to act on some of this. We are in the middle of a time of transition for human society. I really do not know what the world will look like in fifty years' time, but I do know that it will be very different from how it is now. Everyone who is alive during that fifty years has the potential to play a part in the transition to a more benign and peaceful society, living in harmony with the natural world.

My sense is that faith will play a part in this transition, but that it will be a reformed faith. Jesus talked about a deep transition taking place in his society using the phrase, "New wine must be put into new wineskins,"[5] implying that the faith of his day needed a new structure and self-understanding, if it was going to live through the times ahead. I think that is true today. My plea in this book is for a new approach to our inner life that is centered upon shared values and agreed actions and which holds back from defining the divine too closely. In this way a movement for social change could embrace people who are explicitly religious as well as those who are not.

Finally, I think it is helpful to acknowledge that this is a journey for us as individuals and as a society. There is no blueprint that we can suddenly impose. There is no religious experience that will suddenly transform us. Rather we must set out on a journey into the unknown, vulnerable and open to all that we can learn from wherever we might learn it. It is time to rise up like those bones. Whether you are young or one of the older ones, you will find your path. Join hands with others, walk on, through the moments of joy and the moments of doubt, walk on through the valleys of decision and the heights of celebration, walk on and let's make the world a better place for all that lives.

Notes

Chapter One

1. McGilchrist, I. *The Master and His Emissary: the divided brain and the making of the Western world.* Yale University Press: New Haven and London. 2009.
2. Giddens, A. *Modernity and Self-Identity: self and society in the late modern age.* Stanford University Press: Stanford, California. 1991.
3. McGilchrist, I. *The Master and His Emissary.* Phrases extracted describing the left-hemisphere dominated society.
4. Antony, L. with Spence, G. *The Elephant Whisperer: learning about life, loyalty and freedom from a remarkable herd of elephants.* Pan Books, London. 2009 .
5. McGilchrist, I. *The Master and His Emissary.* Chapter 3: 'Language, Truth and Music.'
6. Keens-Soper, A. and Aldous J. (producers). *Animals in Love.* Oxford Scientific Films. First broadcast BBC1, 1 Feb 2015.
7. McGilchrist, I. *The Master and His Emissary.* p. 254.
8. Pinker, S. *The Blank Slate.* Penguin: London. 2002. Appendix on Brown's Universals.
9. Bauckham, R. uses the term 'community of creation' to critique some modern faith positions. Before him it was coined by Aldo Leopold in his extraordinary observations of the natural world detailed in *A Sand County Almanac.*
10. Clare, J., quoted in *Can Poetry Save the Earth?: A Field Guide to Nature Poems* by Felstiner, J. Yale University Press. 2009.
11. Carr, N. *The Shallows: what the internet is doing to our minds.* WW Norton and Company: New York and London. 2010.

Chapter Two

1. Dunbar, R. *Primate Social Systems: studies in behavioural adaptation.* Croom-Helm: London. 1988.

2. Dawkins, R. *The Selfish Gene.* Oxford University Press: Oxford. 1976.
3. Ridley, M. *The Origin of Virtue.* Viking: London. 1996.
4. Midgely, M. *Science and Poetry.* Routledge: Abingdon. 2001.
5. de Waal, F. *The Bonobo and the Atheist: in search of humanism among the primates.* WW Norton & Company: New York. 2013. p. 36.
6. Ibid. p. 50ff.
7. I am specifically not referring to insect social systems here. Their co-operation is primarily instinctive rather than emotional as reflecting a much lesser brain capacity.
8. King, B. *How Animals Grieve.* University of Chicago Press: Chicago and London. 2013.
9. de Waal, F. *Chimpanzee Politics: power and sex among apes.* John Hopkins University Press: Baltimore. 1982; and *Good Natured: the origins of right and wrong in humans and other animals.* Harvard University Press: Cambridge, Massachusetts. 1996.
10. de Waal, F. *The Bonobo and the Atheist.* p. 119.
11. Curtis, A. *The Century of the Self.* British television documentary film. 2002.
12. Pinker, S. *The Blank Slate.* Penguin: London. 2002. p. 330.
13. Sunderland, C. MPhil thesis entitled *Individualism and the breakdown of community life* (Bristol University) 1999. This was consciously trying to understand how the Enlightenment had affected the way that we saw the world.
14. Smith, A. *An Inquiry into the Nature and Causes of the Wealth of Nations.* 1776. Book 1, Chapter 8: 'Of the Wages of Labour'.
15. Standing, G. *The Precariat: the new dangerous class.* Bloomsbury Academic: London and New York. 2011.

Chapter Three

1. Peacocke, A. *Theology for a Scientific Age.* Fortress Press, Minneapolis. 1993.

2. Sample, I. *Massive: the hunt for the God particle.* Virgin Books, London. 2010.
3. It is important to recognize that the word 'material' here is being used as shorthand. Physics, in its latest form, has left corpuscular theories of matter behind. What we perceive as matter is now understood in terms of the interplay between energy and fundamental forces. Most biologists, ecologists and most of the rest of humanity choose to ignore this in our descriptions of life.
4. Voltaire. *Candide.* 1759.
5. Bacon, F. *Novum Organon.* 1620.
6. It should also be noted that science was born as a 'male' pursuit. It would be complicit in much of the gender-stereotyping that dogged the philosophy of the Enlightenment with men seen as rational and constant, while women were perceived as full of passion and of less value.
7. Quoted in Midgley, M. *Science and Poetry.* p. 56ff.
8. Quoted in Atkins, P.W. *Molecular Quantum Mechanics.* Oxford University Press: Oxford. 1983.
9. McGilchrist, I. *The Master and His Emissary.* Describes how the holistic functioning of right-brained processes functions as the true 'master' embracing and directing the 'emissary' which is the left-brained operations of focussed attention. All true creativity arises from the right brain for example, in his view.
10. This process has been developed at an academic level by Frans de Waal, who has dedicated his life to primate psychology. See for example *Our Inner Ape: a leading primatologist explains why we are who we are.* Riverhead Press: New York. 2005.
11. Ceilidh may be known to readers mainly as a dance/music event, but traditionally Ceilidh takes several forms, and storytelling is at its heart.
12. The Chilcott enquiry is to report on this. Its process has been

delayed for many years, the reasons for which are themselves a story.
13. Lyotard, J-F. 'The post-modern condition: a report on knowledge' in *The Theory and History of Literature, Vol 10.* Manchester University Press: Manchester. 1979.
14. Monbiot, G. *Captive State: the corporate takeover of Britain.* Macmillan: Basingstoke and Oxford. 2000. Chapter 9.
15. MacIntyre, A. *After Virtue: a study in moral theory.* Duckworth: London. 1985; and *Whose Justice, Which Rationality.* Duckworth: London. 1988.

Chapter Four

1. Kant, I. *Critique of Pure Reason.* 1781.
2. Haidt, J. *The Righteous Mind.* Allen Lane/Penguin: London and New York. 2012. I am indebted to Jonathan Haidt for much of the analysis in this section.
3. *The Common Cause Handbook.* Public Interest Research Centre. 2011.
4. Ibid.
5. Frazzetto, G. *How We Feel.* Black Swan. 2014. Frazzetto also shows that mice from a strain of poor mothers can become good mothers if they experience nurture by a good mother.
6. Haidt's definition of liberal is from the US and is closer to what we think of as socialist in the UK.
7. McIntyre, A. *After Virtue.* Chapter 15.
8. McGilchrist, I. *The Master and His Emissary.* p. 254.
9. McKnight, J. and Block, P. *Community Abundance Is Its Gifts.* Blog post available at http://www.abundantcommunity.com/home/posts/john_mcknight/parms/1/post/20120426_community_abundance_is_its_gifts.html (accessed February 2016). Also see by the same authors *The Abundant Community: awakening the power of families and neighbourhoods.* Berrett-Koehler Publishers: San Francisco. 2012.

Chapter Five

1. Ward, R. 'Global warming sceptics using media campaign to discredit IPCC'. *The Guardian*. 4 October 2014. Available from http://www.theguardian.com/environment/2013/oct/04/global-warming-sceptics-discredit-ipcc (accessed October 2015).
2. Oreskes, N. and Conway, E.M. *Merchants of Doubt*. Bloomsbury: New York. 2010.
3. Dunlap, R.E. and McCright, A.M. 'Republican and Democratic Views on Climate Change'. *Environment: science policy for sustainable development*. Sept/Oct 2015. Available from http://www.joanneoellers.com/wp-content/uploads/2014/03/Republican-and-Democratic-Views-on-Climate.pdf (accessed Oct 2015).
4. Vaidyanathan, G. 'Big Gap between What Scientists Say and Americans Think about Climate Change'. *Scientific American*. 30 January 2015.
5. Robbins, D. 'Climate Denial Food Chain: Conservative Media Run With Baseless Climate Science Conspiracy Theory'. *Media Matters Research.* 10 February 2015. Available from http://mediamatters.org/research/2015/02/10/climate-denial-food-chain-conservative-media-ru/202469 (accessed Oct 2015).

Chapter Six

1. Ackroyd, P. *Venice*. Vintage Press: London. 2009.
2. Smith, A. *An inquiry in the nature and cause of the wealth of nations*. 1776.
3. Hobbes, T. *Leviathan*. 1651.
4. Mill, J.S. *A System of Logic*. 1858.
5. Hayek, F. *Law, Legislation and Liberty: a new statement of the liberal principles of justice and political economy.* Routledge and Kegan Paul: London. 1979.
6. Daly, H. with Cobb, J.R. *For the Common Good: redirecting the*

economy towards community, the environment and a just and sustainable future. Merlin Press: London. 1989.

7. Dabla-Norris, E., Kochhar, K., Ricka, F., Suphaphiphat, N. and Tsounta, E. (with contributions from Sharma, P. and Salins, V.). 'Causes and Consequences of Income Inequality: A Global Perspective'. *Policy and Review Department.* International Monetary Fund. June 2015. Available at https://www.imf.org/external/pubs/ft/sdn/2015/sdn1513.pdf (accessed February 2016).
8. Wilkinson, R. and Pickett, K. *The Spirit Level: why more equal societies almost always do better.* Allen Lane: London. 2009.
9. Piketty, T. *Capital in the Twenty-First Century.* Harvard University Press. 2013.
10. Macalister, T. and Carrington, D. in the *Guardian* 23 May 2015. Available at http://www.theguardian.com/business/2015/may/22/shell-boss-endorses-warnings-about-fossil-fuels-and-climate-change (accessed October 2015).
11. Neslen, A. 'EU dropped pesticide laws due to US pressure over TTIP, documents reveal'. *The Guardian.* 22 May 2015. Available at http://www.theguardian.com/environment/2015/may/22/eu-dropped-pesticide-laws-due-to-us-pressure-over-ttip-documents-reveal (accessed October 2015).
12. Pettifor, A. *The Coming First World Debt Crisis.* Palgrave: Macmillan. 2006.
13. Anderson, G. *City Boy: beer and loathing in the square mile.* Not Avail Press: London. 2008.
14. Carne, M. 'Inclusive Capitalism: creating a sense of the systemic'. Speech at the *Conference on Inclusive Capitalism.* 27 May 2014. Available at http://www.bankofengland.co.uk/publications/Documents/speeches/2014/speech731.pdf (accessed February 2016).
15. Weber, M. *The Protestant Ethic and the Spirit of Capitalism.* Oxford University Press. 2010 (first published 1920). Chapter 6.

16. Daly, H. with Cobb, J.R. *For the Common Good;* and Klein, N. *This Changes Everything.* Simon and Schuster: New York. 2015.
17. Klein, N. *This Changes Everything.* pp. 67–69.
18. Anderson, G. *City Boy.*
19. Cable, V. *The Storm: the world economic crisis and what it means.* Atlantic Books: London. 2009.

Chapter Seven

1. For a general review on the origins of nations see Hastings, A. *The Construction of Nationhood.* Cambridge University Press: Cambridge, UK. 1997.
2. Curtis, A. *The Bitter Lake.* BBC Documentary 2015. Available at http://www.imdb.com/title/tt4393514/ (accessed February 2016).
3. McKnigh, J. and Block, P. *The Abundant Community.*
4. A summary of research into this question can be found at Hambleton, R. and Sweeting, D. 'The Impacts of Mayoral Governance in Bristol'. *The Bristol Civic Leadership Project.* Available at http://bristolcivicleadership.net/ (accessed October 2015).
5. For example *The Global Parliament of Mayors* project. Details available at http://www.globalparliamentofmayors.org/ (accessed February 2016). This movement began in response to the book by Barber, B. *If Mayors Ruled the World.* Yale University Press. 2013.

Chapter Eight

1. Those who have read Emile Durkheim, for example, and his theories about the power of religious ritual and celebration to bring about social cohesion, may share with me a sense that this analysis does not do justice to the special nature of religion in the human psyche. It is simply too functional and not adequately engaged with an understanding of the

religious mind.

2. Hopkins, G.M. *God's Grandeur*. Available at http://www.poetryfoundation.org/poem/173660 (accessed February 2016).
3. McKibben, B. *The End of Nature*. Random House: New York and London. 2006.
4. RSV version of The Bible, the Book of Job. Chapter 41 extracts from vs 1–8.
5. Macy, J. see *Joanna's Essay on the Council of All Beings*. Available at http://www.joannamacy.net/resources/deepecology/111-joanna-macy-council-of-all-beings-july2002.html (accessed February 2016).
6. An example of scientific studies of this question can be found at Carpenter-Boggs, L., Kennedy A.C. and Reganolda, J.P. 'Organic and Biodynamic Management Effects on Soil Biology' in *Soil Science Society of America Journal*. Vol 64. pp. 1651–59. For a review of all the literature, see Chalker-Scott, L. 'The Science behind Biodynamic Preparations: a literature review' in *Hort Technology*. December 2013. Vol 23. pp. 814–19.
7. *The Nicene Creed*. 1975. Ecumenical Version.
8. I published more than thirty peer-reviewed papers in the field of biochemistry and immunology between 1975 and 1986, having trained at St Peter's College, Oxford.
9. Turkson, P. Address to the *Summit of Consciences for the Climate*. July 2015. Available at http://www.arcworld.org/news.asp?pageID=777 (accessed February 2016).
10. From the artist's point of view, part of the interest in the human body arose not from the faith but from the humanism of the time, yet it remains the case that the patrons were, and are, largely content with this way of proclaiming the faith. It is to be depicted in terms of human beings.
11. Sunderland, C. *The Dream that Inspired the Bible*. EarthAbbey Publishing: Bristol, UK. 2009.

12. Clare, J. taken from *Helpstone.* Available at http://www.johnclare.info/sanada/1Pd2.htm (accessed February 2016).
13. For more information on Permaculture go to *The Permaculture Association.* Available at https://www.permaculture.org.uk/ (accessed February 2016).
14. Pope Francis. *Laudato Si.* Chapter Six. Available at http://w2.vatican.va/content/francesco/en/encyclicals/documents/papa-francesco_20150524_enciclica-laudato-si.html (accessed February 2016).
15. Berry, W. 'Christianity and the survival of creation'. *The Art of the Commonplace: the agrarian essays of Wendell Berry.* Counterpoint Press: Berkeley, CA. 2002. p. 308.
16. Suzuki, D. *The Sacred Balance: rediscovering our place in nature.* Greystone Books: Vancouver. 1997.

Chapter Nine

1. Intergovernmental Panel on Climate Change reports. Available at http://www.ipcc.ch/ (accessed February 2016).
2. This is not to ignore the many serious thinkers for whom our relationship with the Earth has been a long-term project and whose history goes back to the 1970s and before. The last ten years have been the time when governments and the people en masse have accepted that there is a problem.
3. 'Latest trends in public attitudes to taking action on climate change'. *UN Climate Change Conference: hearing the public's voice.* Ipsos MORI Social Research Institute. December 2015.
4. Greco, T. *The End of Money and the Future of Civilisation.* Chelsea Green Publishing: Vermont. 2009.
5. Monbiot, G. *Heat.* Penguin: London. 2006.
6. The local multiplier effect has been researched by the New Economics Foundation. They show that money that passes through several hands before it leaves a community is of more value to the local economy than money that goes straight to the financial institutions. There is a video about it

available at http://youtu.be/cj3Bp13hLiE (accessed February 2016).

Chapter Ten

1. James, C. 'Japanese Maple'. Available at http://www.dying-matters.org/page/clive-james-japanese-maple (accessed February 2016).
2. Bate, J. *John Clare: a biography.* Farrar, Strauss and Giroux: New York. 2005.
3. Clare, J. 'Beans in Blossom'. Section XXVII. Available at http://www.johnclare.info/sanada/4Rm3.htm (accessed February 2016).
4. The Hebrew words translated as breath and spirit are the same.
5. Psalm 104. RSV version of The Bible, vs 29–30.
6. The Book of Isaiah. RSV version of The Bible. Chapter 40, v 31.
7. The Gospels of Luke. The Bible. Chapter 15, vs 11ff; and Matthew. The Bible. Chapter 25, vs 31ff,
8. Newbiggin, L. *Foolishness to the Greeks.* Eerdmans, London. 1986.
9. See for example the positive impacts on civilization noted in Stark, R. *For the Glory of God.* Princeton University Press. 2004.

Chapter Eleven

1. Sunderland, C.A. *Individualism and the Breakdown of Community Life.* MPhil dissertation. Trinity College Bristol, validated by University of Bristol. 1999.
2. Fukuyama, F. *The End of History and the Last Man.* The Free Press: New York. 1992.
3. Durkheim, E. *The Elementary Forms of the Religious Life.* 1912. Durkheim saw that ritual and celebration lay at the very heart of the energy in the world's great religions. People were

inspired by the very act of meeting and participation in a ritual.

4. Please note that this is quite different in approach from the Sacred Lands project which is about historical, religious land use.

Chapter Twelve

1. The Book of Ezekiel. The Bible. Chapter 37.
2. McKnight, J. and Block, P. *Community Abundance Is Its Gifts.* Blog post. Available at http://www.abundantcommunity.com/home/posts/john_mcknight/parms/1/post/20120426_community_abundance_is_its_gifts.html (accessed February 2016). Also see by the same authors *The Abundant Community.*
3. McGilchrist, I. *The Master and His Emissary.*
4. A fuller description of this method of analysis is available at http://valuesandframes.org/handbook/2-how-values-work/ (accessed February 2016).
5. The Gospel of Luke. The Bible. Chapter 5, vs 36–39.

Bibliography

Ackroyd, P. *Venice.* Vintage Press: London. 2009.

Anderson, G. *City Boy: beer and loathing in the square mile.* Not Avail Press: London. 2008.

Antony, L. with Spence, G. *The Elephant Whisperer: learning about life, loyalty and freedom from a remarkable herd of elephants.* Pan Books: London. 2009.

Atkins, P.W. *Molecular Quantum Mechanics.* Oxford University Press: Oxford. 1983.

Bacon, F. *Novum Organon.* 1620.

Bate, J. *John Clare: a biography.* Farrar, Strauss and Giroux: New York. 2005.

Berry, W. *The Art of the Commonplace: the agrarian essays of Wendell Berry.* Counterpoint Press: Berkeley, CA. 2002.

Carr, N. *The Shallows: what the internet is doing to our minds.* WW Norton & Company: New York and London. 2010.

Cable, V. *The Storm: the world economic crisis and what it means.* Atlantic Books: London. 2009.

Common Cause Handbook. Public Interest Research Centre. 2011.

Curtis, A. *The Century of the Self.* British television documentary film. 2002.

Daly, H. with Cobb, J.R. *For the Common Good: redirecting the economy towards community, the environment and a just and sustainable future.* Merlin Press: London. 1989.

Dawkins, R. *The Selfish Gene.* Oxford University Press: Oxford. 1976.

de Waal, F. *Chimpanzee Politics: power and sex among apes.* John Hopkins University Press, Baltimore. 1982.

- - -. *Good Natured: the origins of right and wrong in humans and other animals.* Harvard University Press: Cambridge, Massachusetts. 1996.

- - -. *Our Inner Ape: a leading primatologist explains why we are who we are.* Riverhead Press: New York. 2005.

- - -. *The Bonobo and the Atheist: in search of humanism among the primates.* WW Norton & Company: New York. 2013.

Dunbar, R. *Primate Social Systems: studies in behavioural adaptation.* Croom-Helm: London. 1988.

Durkheim, E. *The Elementary Forms of the Religious Life.* 1912.

Frazzetto, G. *How We Feel.* Black Swan. 2014.

Fukuyama, F. *The End of History and the Last Man.* The Free Press: New York. 1992.

Giddens, A. *Modernity and Self-Identity: self and society in the late modern age.* Stanford University Press: Stanford, California. 1991.

Greco, T. *The End of Money and the Future of Civilisation.* Chelsea Green Publishing: Vermont. 2009.

Haidt, J. *The Righteous Mind.* Allen Lane/Penguin: London and New York. 2012.

Hastings, A. *The Construction of Nationhood.* Cambridge University Press: Cambridge, UK. 1997.

Hayek, F. *Law, Legislation and Liberty: a new statement of the liberal principles of justice and political economy.* Routledge and Kegan Paul: London. 1979.

Hobbes, T. *Leviathan.* 1651.

Kant, I. *Critique of Pure Reason.* 1781.

King, B. *How Animals Grieve.* University of Chicago Press: Chicago and London. 2013.

Klein, N. *This Changes Everything.* Simon and Schuster: New York. 2015.

Lyotard, J-F. 'The post-modern condition: a report on knowledge' in *The Theory and History of Literature.* Vol 10. Manchester University Press: Manchester. 1979.

MacIntyre, A. *After Virtue: a study in moral theory.* Duckworth: London. 1985.

- - -. *Whose Justice, Which Rationality.* Duckworth: London. 1988.

McGilchrist, I. *The Master and His Emissary: the divided brain and the making of the Western world.* Yale University Press: New Haven and London. 2009.

McKibben, B. *The End of Nature.* Random House: New York and London. 2006.

McKnight, J. and Block, P. *The Abundant Community: awakening the power of families and neighbourhoods.* Berrett-Koehler Publishers: San Francisco. 2012.

Midgely, M. *Science and Poetry.* Routledge: Abingdon. 2001.

Mill, J.S. *A System of Logic.* 1858.

Monbiot, G. *Captive State: the corporate takeover of Britain.* Chapter 9. Macmillan: Basingstoke and Oxford. 2000.

- - -. *Heat.* Penguin: London. 2006.

Newbiggin, L. *Foolishness to the Greeks.* Eerdmans: London. 1986.

Oreskes, N. and Conway, E.M. *Merchants of Doubt.* Bloomsbury: New York. 2010.

Peacocke, A. *Theology for a Scientific Age.* Fortress Press: Minneapolis. 1993.

Pettifor, A. *The Coming First World Debt Crisis.* Palgrave Macmillan: London. 2006.

Piketty, T. *Capital in the Twenty-First Century.* Harvard University Press. 2013.

Pinker, S. *The Blank Slate.* Penguin: London. 2002.

Ridley, M. *The Origin of Virtue.* Viking: London. 1996.

Sample, I. *Massive: the hunt for the God particle.* Virgin Books: London. 2010.

Smith, A. *An Inquiry into the Nature and Causes of the Wealth of Nations.* 1776.

Standing, G. *The Precariat: the new dangerous class.* Bloomsbury Academic: London and New York. 2011.

Stark, R. *For the Glory of God.* Princeton University Press. 2004.

Sunderland, C. *Individualism and the Breakdown of Community Life.* MPhil dissertation. Trinity College Bristol, validated by University of Bristol. 1999.

- - -. *The Dream that Inspired the Bible.* EarthAbbey Publishing: Bristol. 2009. (Available from the author.)

Suzuki, D. *The Sacred Balance: rediscovering our place in nature.* Greystone Books: Vancouver. 1997.

Voltaire. *Candide.* 1759.

Weber, M. *The Protestant Ethic and the Spirit of Capitalism.* Oxford University Press. 2010 (first published 1920).

Wilkinson, R. and Pickett, K. *The Spirit Level: why more equal societies almost always do better.* Allen Lane: London. 2009.

About the Author

Chris Sunderland has been, as they say, many things to many people. He has been a research biochemist, working in an Oxford laboratory at the cutting edge of biotechnology. He has been a vicar in one of the toughest tower-block estates in the UK. He has become a storyteller engaging with issues of public life. He is one of the founding directors of the Bristol Pound, a local currency project that has attracted widespread attention and is currently developing a new food cooperative, called Real Economy, designed to get fresh food to people in the toughest circumstances.

Chris is convinced that we are living through an age of deep change, when all of our cultural systems, including our politics, our economics and our religious institutions, are in decline and something new is emerging. At this moment the one thing that human beings must pay attention to is our relationship with the Earth.

The Bristol Pound and its sister project Real Economy are an attempt to give people a taste of a different type of economic system, which is led by values, full of relationships and a driver of change towards a sustainable culture.

Chris blogs at the site withwingslikeeagles.org and can be found on Twitter @wwingsleagles.

EARTH

BOOKS

Earth Books
ENVIRONMENT

Earth Books are practical, scientific and philosophical publications about our relationship with the environment. Earth Books explore sustainable ways of living; including green parenting, gardening, cooking and natural building. They also look at ecology, conservation and aspects of environmental science, including green energy. An understanding of the interdependence of all living things is central to Earth Books, and therefore consideration of our relationship with other animals is important. Animal welfare is explored. The purpose of Earth Books is to deepen our understanding of the environment and our role within it. The books featured under this imprint will both present thought-provoking questions and offer practical solutions.
If you have enjoyed this book, why not tell other readers by posting a review on your preferred book site. Recent bestsellers from Earth Books are:

In Defence of Life

Essays on a Radical Reworking of Green Wisdom

Julian Rose

Julian Rose's book has the power to lift the reader into another dimension. He offers a way to break through the destructive patterns of our consumer-obsessed society and discover a simpler, more fulfilling way forward.

Paperback: 978-1-78279-257-4 ebook: 978-1-78279-256-7

Eyes of the Wild

Journeys of Transformation with the Animal Powers

Eleanor O'Hanlon

The ancient understanding of animals as guides to self-knowledge and the soul comes alive through close encounters with some of the most magnificent creatures of the wild.

Paperback: 978-1-84694-957-9 ebook: 978-1-84694-958-6

Simplicity Made Easy

Jennifer Kavanagh

Stop wishing your life was more simple, and start making it happen! With the help of Jennifer Kavanagh's book, turn your focus to what really matters in life.

Paperback: 978-1-84694-543-4 ebook: 978-1-84694-895-4

Acorns Among the Grass

Adventures in Eco-Therapy

Caroline Brazier

When we reconnect with the natural world, we discover our deep relationship with life. This book embraces the ways in which environmental work nourishes us, psychologically and spiritually.

Paperback: 978-1-84694-619-6 ebook: 978-1-84694-883-1

Safe Planet

Renewable Energy Plus Workers' Power

John Cowsill

Safe Planet lays out a roadmap of renewable energy sources and meteorological data to direct us towards a safe planet.

Paperback: 978-1-78099-682-0 ebook: 978-1-78099-683-7

Approaching Chaos

Could an Ancient Archetype Save 21st Century Civilization?

Lucy Wyatt

Civilisation can survive by learning from the social, spiritual and technological secrets of ancient civilisations such as Egypt.

Paperback: 978-1-84694-255-6

Gardening with the Moon & Stars

Elen Sentier

Organics with Ooomph! Bringing biodynamics to the ordinary gardener.

Paperback: 978-1-78279-984-9 ebook: 978-1-78279-985-6

GreenSpirit

Path to a New Consciousness

Marian Van Eyk McCain

A collection of essays on 21st Century green spirituality and its key role in creating a peaceful and sustainable world.

Paperback: 978-1-84694-290-7 ebook: 978-1-78099-186-3

The Protein Myth

Significantly Reducing the Risk of Cancer, Heart Disease, Stroke, and Diabetes While Saving the Animals and the Planet

David Gerow Irving

The Protein Myth powerfully illustrates how the way to vibrant health and a peaceful world is to stop exploiting animals.

Paperback: 978-1-84694-673-8 ebook: 978-1-78099-073-6

This Is Hope
Green Vegans and the New Human Ecology How We Find Our Way to a Humane and Environmentally Sane Future
Will Anderson
This Is Hope compares the outcomes of two human ecologies; one is tragic, the other full of promise...
Paperback: 978-1-78099-890-9